'We have needed this book for many years but the Climate Emergency and the growing popular unrest among students and young professionals makes this comprehensive approach a must-read book for the whole industry'.
  – *Robin Nicholson CBE, RIBA, Hon FIStructE, Hon FCIBSE, Convenor of the Edge and a long-standing Partner of Cullinan Studio*

'Across the globe, climate emergency declarations cover more than 1 billion people in 2,000 jurisdictions and local governments. Many of these overlay our urban areas where the built environment is an important contributor to carbon emissions. Retrofitting the 70% of buildings that will still be here in 2050 is crucial to tackling the climate emergency, creating a resilient built environment, and bending the curve to keep temperature rise to within 1.5 degrees. This impressive, international, multi-scale and interdisciplinary-based book shows how the many and varied buildings that make up the built environment can be made more energy, waste, and water efficient and tackles vitally important questions: the "how", the "what", and the "why" of retrofitting. The authors have written an eloquent and authoritative guide to these fundamental issues, and this is underpinned by a passionate manifesto, or call to arms, for professionals of all disciplines to tackle these issues right now and at scale. Finally, the book is an especially poignant testament to the late Sarah Sayce's excellent and well-respected work in this field'.
  – *Professor Timothy J. Dixon, School of the Built Environment, University of Reading*

Resilient Building Retrofits

This radical book aims to inject new insight and urgency into the discourse on the retrofitting of commercial and residential buildings in the face of the climate emergency. It is about the why, how and who should take the lead in revolutionising buildings in the face of serious climate and social change.

Buildings contribute very significantly to the output of carbon, particularly in developed countries where the stock is old, but it is neither feasible nor desirable to demolish them all and start again! If existing buildings cannot be replaced in the short term by new zero-carbon stock, retrofitting and adaptation of the existing building stock is critical and urgent. This book explains why and how the improvement of buildings requires a complex, holistic approach that brings all stakeholders together with respect and understanding. Yet to do this against a limited time frame is challenging. The book analyses what must be done, explores how it could be achieved and sets out a manifesto for action by all those engaged: from policymakers to educationalists, designers, constructors, investors, funders and occupiers.

By bringing together authors from across the built environment disciplines, the book stimulates debate within policy, practice and education circles which must lead to action if we are to avoid catastrophe. This is a unique addition to the literature on the sustainability of existing buildings and their retrofitting for the benefit of all.

Sarah Sayce was Professor of Sustainable Real Estate at Henley Business School, University of Reading, and Visiting Professor at the Royal Agricultural University, UK. For many years she headed the School of Surveying and Planning at Kingston University, UK. She recently worked on two EU-funded projects in the field of energy efficiency and property values. She is the co-author of *Developing Property Sustainably* and co-editor of the *Routledge Handbook of Sustainable Real Estate*, both published by Routledge.

Sara Wilkinson is Professor in the School of Built Environment at the University of Technology Sydney, Australia. Her research focuses on sustainability and adaptation in the built environment, user satisfaction, retrofit of green roofs and conceptual understanding of sustainability. Sara is on the editorial board of five international refereed journals. She is the co-author of *Developing Property Sustainably* and co-editor of the *Routledge Handbook of Sustainable Real Estate*, both published by Routledge.

Gillian Armstrong is a chartered architectural technologist and a UK registered architect. Gillian is a post-doctoral researcher at University Technology Sydney. Her research crosses architecture, surveying, and planning disciplines and focuses on building regulation, and sustainable design and management of existing buildings. Gill has also published on built environment pedagogies and crisis narratives in education, having taught university built environment programmes in Australia and the UK.

Samantha Organ is a chartered surveyor leading nationally on sustainability for the built estate for Europe's largest conservation charity, the National Trust. She is the MSc Programme Leader and Senior Lecturer in Building Surveying at the University of the West of England, Bristol. Her research focuses primarily on sustainability in the built environment. Sam has written numerous publications for a range of audiences and is a reviewer for a number of peer-reviewed academic journals. She won the RICS Matrics Young Building Surveyor of the Year in 2020.

Resilient Building Retrofits

Combating the Climate Crisis

**Edited by Sarah Sayce, Sara Wilkinson,
Gillian Armstrong and Samantha Organ**

LONDON AND NEW YORK

Cover image: FilippoBacci

First published 2023
by Routledge
4 Park Square, Milton Park, Abingdon, Oxon OX14 4RN

and by Routledge
605 Third Avenue, New York, NY 10158

Routledge is an imprint of the Taylor & Francis Group, an informa business

© 2023 selection and editorial matter, Sarah Sayce, Sara Wilkinson, Gillian Armstrong and Samantha Organ; individual chapters, the contributors

The right of Sarah Sayce, Sara Wilkinson, Gillian Armstrong and Samantha Organ to be identified as the authors of the editorial material, and of the authors for their individual chapters, has been asserted in accordance with sections 77 and 78 of the Copyright, Designs and Patents Act 1988.

British Library Cataloguing-in-Publication Data
A catalogue record for this book is available from the British Library

Library of Congress Cataloging-in-Publication Data
A catalog record for this book has been requested

ISBN: 978-0-367-90355-8 (hbk)
ISBN: 978-0-367-90354-1 (pbk)
ISBN: 978-1-003-02397-5 (ebk)

DOI: 10.1201/9781003023975

Typeset in Goudy
by Apex CoVantage, LLC

Contents

Contributors

Ursula Hartenbeger
Director, PathTo2050, Brussels, Belgium

Professor Jeroen van der Heijden
Wellington School of Business and Government | Ōrauariki
Te Herenga Waka – Victoria University of Wellington, New Zealand

Dr Shabnam Yazdani Mehr
GBA Heritage, Sydney, NSW Australia
University of Technology Sydney, Sydney, NSW, Australia

Associate Professor Hilde Remoy
Technical University of Delft
Rotterdam, the Netherlands

Zsolt Toth
Buildings Performance Institute Europe (BPIE), Brussels, Belgium

Preface

This is a message for everyone, especially professionals and young people. The 2021 COP conference in Glasgow highlighted the need to act globally; it is probably too early to see what it has achieved, but it will need the commitment of everybody to make the necessary real shift. Hopefully, this book will give some inspiration to make that shift. Some of the sources cited go a long way back, so it is not a new message but one that some have chosen not to hear. I hope this book will make them hear.

To those who have power, or some sort of power under their control, please listen and take heed of the main message in this book. I urge you not to come out with platitudes merely perhaps to respond to those like Greta Thunberg; what she and others said repeatedly is to read and abide by the science, and do not let national and personal motives get in the way. The world's economy is important; it keeps the world turning, but the planet is more important arguably. At last we seem to be marrying the human motivation of moving forward with the impact on the planet, but do let the latter frame and mould the former.

Please, on my behalf, and on behalf of my fellow contributors, it matters; it really does. To all property professionals, it is too easy to sit on your hands and say 'let's set up a working party', 'we are going to do things'. But that is not enough. We do not have the luxury of years ahead anymore. So act on your convictions for all our sakes and those of the coming generation.

My many thanks to Sara for the way she has driven, supported and helped. We worked closely together to complete this book. So also my thanks in particular to all our co-authors who gave their time so freely and to their dedication to contributing to everything that sits within this book. And a special thanks to Sam and Gill for stepping in at this late moment to do some of that heavy lifting at the end with the editing and final chapter writing in order to ensure with Sara that the book weaves into a narrative that makes sense in so many ways.

Sarah Sayce
18 November 2021

This was dictated just a few days before Sarah passed away with cancer; she was very keen that action should occur, not just talking about it, and she devoted many years to this area trying to get property in all its guises to be more friendly to our environment and also for all to recognise the significant contribution that property has made to the challenge before us, and how we can manage and use property better and more sustainably. Particularly, it was changing people's minds and educating them to work for a truly sustainable future which drove her.

Part 1

The why

Challenge and a need to change

1 The climate crisis

Why it matters for the built environment

Sarah Sayce, Sara Wilkinson and Samantha Organ

1.1 Introduction

No book that deals with the matter of buildings can afford to ignore the escalating issue of climate change and its potentially disastrous environmental, economic and social consequences. It is widely acknowledged that the whole life cycle of a building, from its location through to its construction, use, alteration and ultimate destruction, has a major influence on the climate, primarily, though not exclusively, through its contribution to carbon emission (Clayton et al., 2021). If the principles of the circular economy (Geissdoerfer et al., 2017; Kirchherr et al., 2017) are accepted, then some key factors for buildings can be elicited as critical:

1. Initial resource use should be minimised, allowing maximum reuse of materials.
2. The building should be designed to support minimal (or no) carbon use in construction and operation.
3. The building life should be maximised through low-resource intensity retrofits to sustain its longevity for the optimal period.
4. Demolitions should be done in a way that circularity can be achieved through material reuse and, where this is not possible, recycling.

This book deals primarily with the third point: the importance of the retrofit process in lengthening the building life, reducing operational carbon use and supporting social change – all within an economic framework that supports, rather than denies, such processes.

This may sound overambitious, even idealistic, but, we argue, delivery on this is 'mission critical' to any realistic possibility of achieving the targets for zero-carbon economies that are being progressively set out and accepted by many corporates, institutions and governments around the globe. In so doing, we are acutely aware that even accepting that climate change is anything more than a natural phenomenon is not universal. However, as the Paris Agreement, referred to as the Paris Accord (UNFCCC, 2020), testifies, the prevailing majority view is that it is real, it has a strong anthropogenic component and the response must be one of both mitigating future changes to stay with global temperature rises of 1.5 degrees Celsius and of adapting to what is already inevitable and locked in.

DOI: 10.1201/9781003023975-2

The urgent need to accept that human actions are resulting in climate change and that it is possible to take effective behaviour change action, both collective and individual, has proved extraordinarily difficult. Even now, as the Covid-19 health pandemic takes its toll globally, the reluctance to accept that it will not be a return to 'business as usual' is proving to be an unpalatable message, primarily for economic reasons. The Covid-19 pandemic is something that is portrayed vividly with images of ventilators and human suffering in every country. It is far harder to engender and maintain actions relating to climate change that, for most, seem distant – or transient (Clayton et al., 2015). In December 2021, the World Meteorological Organization reported a new record temperature of 38 degrees Celsius in the Arctic (WMO, 2021); whilst the scenes of wildfires in Australia and California created shocking news images in 2019 and early 2020, the concept of availability heuristics (Mase et al., 2015) instilled collective myopia unless the image was constantly made vivid and relevant. It is the mindset so produced that Ord (2020) means that we are not adept at anticipating and acting on catastrophes for which we do not have precedents; therefore, he argues, even when a threat is very real and clearly explained by scientists, we have great difficulty in believing it will happen until it does. Ord (2020) argues that climate change is one, but only one, such existential threat.

Through the chapters that follow, we lay down the challenge of what it will take to make buildings capable, so supporting the moves to a zero-carbon global society in which temperature increase can be capped to 1.5 or, at worst, 2 degrees Celsius. The challenge will require concerted, sustained and collaborative efforts by all stakeholders: policymakers, building owners, building occupiers, the design and construction industry and financiers. Some 15 years ago, Stern (Stern & Stern, 2007) predicted that the financial costs of not dealing with the matter of climate change immediately would be enormous.

Using the results from formal economic models, the Review estimates that if we do not act, the overall costs and risks of climate change will be equivalent to losing at least 5% of global GDP each year, now and forever. If a wider range of risks and impacts is taken into account, the estimates of damage could rise to 20% of GDP or more (Stern & Stern, 2007).

Alternatively, if mitigation and adaptation measures are adopted, much of this economic downside could be lessened, though not totally avoided, to around 1% of global GDP (Stern & Stern, 2007). However, opinions were divided on the validity of conclusions drawn. Eventually, the Paris Agreement, an international treaty on climate change, was adopted in 2015. It covers climate change mitigation, adaptation and finance.

Regretfully, such warnings and the science from which the economic models were built were not heeded, and since that time, both the severity of the issue and the potential costs have escalated – and continue to escalate to the point when 33 countries and 1,863 jurisdictions/cities have recognised and declared a climate crisis (Climate Emergency Declaration, 2020). However, the debates continue to this day, and the most recent UN Climate Change Conference, also referred to as the COP26, took place in Glasgow in November 2021. Recent weather events have placed extreme weather and its social, economic and environmental impacts at the forefront of many people's minds. Examples of these events are the 2021 wildfires in Canada, the 2020 bushfires and droughts in Australia, and the 2021 floods in Germany. Though the nature of the crisis and its physical manifestations are complex, they are explored within this introductory chapter.

The climate crisis is not reserved for economic consequences. They may be almost unimaginable, but the social costs of dislocated communities and the attendant poverty and sickness and premature death are far, far higher (Tol, 2002). Such costs will fall, inevitably, upon poorer nations and poorer communities, whilst those living in richer states and countries may have the ability to buy their way out of the immediate impacts, at least in the short term (Stern & Stern, 2007). However, not completely, as the changing climate presents risks that are now feeding through into regulatory frameworks, which will doubtless impact individual and collective freedoms.

This chapter sets out some of the latest thinking on climate change scenarios, including information from the Intergovernmental Panel on Climate Change (IPCC), which points to the inevitable and potential physical consequences: from the impact of rising land temperatures to rising sea levels, floods, droughts, fires and storms (IPPC, 2007). It explores the social consequences of species extinction, disruption to food chains and mass migration. Collectively, these potential consequences, if they come to pass, present an existential risk.

Collectively, these events present challenges to every level of the built environment, from city design to transport infrastructure and, the subject of this book, to the level of how we can adapt and use individual buildings to help mitigate some of the most severe consequences. In so doing, it accepts that mitigation of climate change, whilst laudable, and to date, the preferred policy approach is an inadequate response and requires greater thinking and action around adaptation. In spelling out the challenges, this chapter sets the scene for the book.

Before detailing the size of the challenge, it is important to spell out the contribution that buildings make to carbon emissions. Estimates vary, but most agree that they represent around 40% of emissions (Wilby, 2007). Some of this relates to construction – but much is as a result of the way that buildings are accessed, operated, managed and used, and, at the end of their life cycle, demolished.

Before going on to detail what the latest thinking is on climate change scenarios, it is important to understand the types of carbon emission, known as Scopes 1, 2 and 3. Scope 1 emissions are produced from manufacturing processes, such as from cement manufacture, and emissions from the burning of diesel fuel in trucks and fugitive emissions, such as methane emissions from coal mines, or the production of electricity by burning coal. Another summary is Scope 1 covers direct emissions from owned or controlled sources. Scope 2 emissions are indirect emissions from the generation of purchased energy from a utility provider. In other words, it includes all greenhouse gas (GHG) emissions released into the atmosphere from the consumption of purchased electricity, steam, heat and cooling. Scope 3 includes all other indirect emissions that occur in a company's value chain. They are a consequence of the activities of the company but occur from sources not owned or controlled by the company. Some examples of Scope 3 activities are extraction and production of purchased materials, transportation of purchased fuels and use of products and services.

1.2 The climate crisis

In a 2020 report, Bill Gates and Melinda French Gates stated, *'There is no such thing as a national solution to a global crisis'*. Said in relation to the Covid-19 pandemic, it holds true for climate change. Like the virus, climate change does not adhere to borders, and neither should our solutions. The climate crisis affects us all.

Recent climate changes have been widespread and are affecting every region. Although different regions are experiencing different combinations of changes and impacts, the changes are unprecedented, intensifying and becoming more frequent. As global warming increases, the changes in the climate become larger. Evidence suggests that climate extremes such as heatwaves, record-high temperatures, disruptive precipitation patterns, wildfires (Lyster, 2017), sea-level rise and ocean acidification (Whyte, 2019), and droughts and floods are all consequences of anthropogenic climate change (Ide et al., 2020). Such events pose a threat to human security, food chains, health and disease, access to water, biodiversity (Ide et al., 2020) as well as our cities (Lyster, 2017). A number of these changes are unprecedented, including the rate of sea-level rise and glacial retreat, and irreversible in our lifetimes. This means there is a need to implement adaptation and mitigation strategies, as well as increased efforts to slow irreversible changes.

a) IPCC climate change scenarios

The IPCC report (Rogelj et al., 2019) set out mitigation pathways compatible with 1.5 degrees Celsius temperature increase in the context of sustainable development. Limiting warming to 1.5 degrees Celsius will mean reaching net zero carbon emissions globally. To achieve this, mitigation pathways under the IPCC incorporate energy demand reductions, the decarbonisation of electricity and other fuel sources and some form of carbon storage or sequestration. Average global temperatures are already 1.1 degrees Celsius warmer than temperatures in 1850–1900. The world is currently projected to achieve a global temperature increase of 2.7 degrees Celsius, reduced to 2.4 degrees Celsius if all the Nationally Determined Contributions under the Paris Accord are achieved. This would have a monumental impact.

In their 2021 report, the IPCC modelled five shared socio-economic pathways (SSPs). These used low, intermediate, high and very high greenhouse gas scenarios to predict the impact on global warming (Table 1.1). Under these scenarios, four of the SSPs saw a global temperature increase in excess of 1.5 degrees Celsius relative to 1850–1900 temperatures, only the very low GHG scenario (SSP1–1.9) predicting a 1.4 degrees Celsius increase as the 'best estimate' in the long term, but with an overshoot to 1.9 degrees Celsius in the medium term. For the pathways that temporarily overshoot the 1.5 degrees Celsius scenario, large-scale carbon

Table 1.1 Changes in global surface temperature adapted from IPCC (2021)

Scenario	Greenhouse gas levels	Near term (2021–2040)		Medium term (2041–2060)		Long term (2081–2100)	
		Best estimate (degrees Celsius)	Very likely range (degrees Celsius)	Best estimate (degrees Celsius)	Very likely range (degrees Celsius)	Best estimate (degrees Celsius)	Very likely range (degrees Celsius)
SSP1–1.9	Very low	1.5	1.2–1.7	1.6	1.2–2.0	1.4	1.0–1.8
SSP1–2.6	Low	1.5	1.2–1.8	1.7	1.3–2.2	1.8	1.3–2.4
SSP2–4.5	Intermediate	1.5	1.2–1.8	2.0	1.6–2.5	2.7	2.1–3.5
SSP3–7.0	High	1.5	1.2–1.8	2.1	1.7–2.6	3.6	2.8–4.6
SSP5–8.5	Very high	1.6	1.3–1.9	2.4	1.9–3.0	4.4	3.3–5.7

emissions removal measures are relied on, but these currently remain unproven at scale and therefore represent significant uncertainty and risk.

Even the 'low' scenario (SSP1–2.6) is likely to see a temperature increase of 1.8 degrees Celsius. In contrast, for the 'very high' scenario (SSP5–8.5), temperatures could increase by 4.4 degrees Celsius, but as much as 5.7 degrees Celsius at the top of the range. Such temperature increases would have catastrophic effects.

There are a number of barriers to achieving these SSPs. These include a lack of global co-operation, a lack of governance in relation to energy and land trans-formation and increased resource-intensive consumption (Shukla et al., 2019).

b) The socio-political consequences

There are a number of socio-political consequences arising from climate change. It is now accepted that some negative effects from climate change will be unavoid-able. The societal and financial consequences of anthropogenic climate change have become known as 'loss and damage', which was included in a clause in the Paris Accord (COP21). 'Loss' refers to an irreversible change, and 'damage' suggests the potential for repair. However, this could be a matter of life and death for some nations, with countries such as Kenya already experiencing drought, crop failure and starvation, and other nations such as Tuvalu experiencing sea-level rises.

In the context of the adverse effects of climate change, it is those disadvantaged groups that have been shown to suffer disproportionately. This not only leads to greater social inequalities, but it can also increase the susceptibility to the dam-age resulting from climate change and reduce the disadvantaged group's ability to recover from such damage (Islam & Winkel, 2017). However, great care is needed when designing interventions. Sources of vulnerability can be reinforced, redis-tributed or created through interventions aimed at climate change adaptation and reductions in vulnerability (Eriksen et al., 2021). Such 'maladaptive outcomes' can be driven by what Eriksen et al. (2021) describe as a 'shallow' understanding of the 'vulnerability context', inequitable stakeholder participation, adding adaptation to existing development agendas and a lack of critical engagement on defining success.

Climate change has been identified by the IPPC as having the potential to become a significant contributory factor in exacerbating the scarcity of natural resources such as freshwater (Bernstein et al., 2007). Shukla et al. (2019) suggest that limiting global warming to 1.5 degrees Celsius compared to 2 degrees Celsius may reduce the proportion of the world's population exposed to increased water stress resulting from climate change. Although it is estimated that this reduction can be as much as 50%, this can vary significantly between regions.

c) *Species extinction*

Humans depend on a broad and complex range of ecosystems. Species extinction represents a threat to food security and health, with a loss of diversity in our diets linked to health risk factors (UN Action, 2019). Healthy forests and woodlands clean and cool the air; mangroves provide protection to tropical coastlines against storm surge damage. Changes to these ecosystems will have a profound effect on society (Ide et al., 2020). Protecting and restoring nature is beneficial for people and climate change.

Nature is not only essential for human life, but it is also recognised as important for the *quality* of life (Intergovernmental Science-Policy Platform on Biodiversity and Ecosystem Services, IPBES, 2019). Globally, biodiversity is declining at an unprecedented rate, and the rate of species extinction is accelerating (IPBES, 2019). Although climate change is only one pressure on our wildlife, it is recognised as playing an important role. Indeed, without careful management, rapid climate change is likely to contribute to the extinction of critically endangered species and to other species becoming even rarer. Bernstein et al. (2007) estimated that globally a quarter of mammals and 12% of birds are at significant risk of extinction. Carbon emissions have led to increases in seawater acidity, known as 'ocean acidification'. Climate change–driven ocean acidification is resulting in rapid changes to global ecosystems and a threat to marine life (Greenhill et al., 2020). Biodiversity loss cannot be resolved without addressing climate change, and climate change cannot be resolved without addressing biodiversity.

The need to protect and restore nature and ecosystems, including forests, was incorporated under the Glasgow Climate Pact at COP26. This inclusion is fundamental and complementary in limiting global temperature rises to 1.5 degrees Celsius. Through COP26, although over 130 countries have pledged to halt and reverse forest loss and land degradation by 2030, previous COP pledges to reverse biodiversity loss have been largely unfulfilled, and no mechanism was in place to enforce such pledges.

d) *Disruption to food chains*

Agriculture, forestry and land use globally account for 23% of anthropogenic greenhouse gas emissions (IPCC, 2020). However, this sector also supports global food security and millions of jobs. For example, in India, the agricultural industry employs over half of the workforce (Guntukula, 2019). Changes to our climate will impact our food chains, and there is a need to adapt our agriculture and food systems both nationally and globally, particularly in relation to food security, global population growth, political turbulence and shocks such as the Covid-19 pandemic (Wheeler & Lobley, 2021).

For some regions, the impacts resulting from climatic changes are not all negative. For example, in the UK, it has led to the opportunity to introduce crops such as grapes that were previously not viable due to local climatic conditions (Wheeler & Lobley, 2021). Some lower latitude regions have seen reduced yields in crops such as maize and wheat (IPCC, 2020), whereas higher latitude regions have experienced greater yields of these crops (IPCC, 2020), resulting from extreme temperatures (Wheeler & Lobley, 2021; Guntukula, 2019; Ayanlade et al., 2017) and heavy rain (Wheeler & Lobley, 2021; Guntukula, 2019). Agricultural impacts are not reserved to crop production – livestock could be impacted by changes in the climate, including lower availability of quality food for the livestock, lower livestock weights, lower fertility and increased livestock mortality (Wheeler & Lobley, 2021).

In addition to such stresses, the IPCC (2020) also reported increases and decreases in agricultural pests and diseases. Such strain on global food production capacity is likely to have an effect on global food availability. For those with limited resources, their climate change resilience is likely to be limited. In their study based on Nigerian farmers, Ayanlade et al. (2017) found that due to their limited resources, smallholder farmers are particularly vulnerable to climate change, limiting their ability to be climate resilient.

Analyses by the IPCC (Rogelj et al., 2019) have indicated that when limiting global temperature increases to 1.5 degrees Celsius compared to 2 degrees Celsius rise, smaller net reductions in crop yields were achieved, particularly in sub-Saharan Africa, Southeast Asia and Central and South America. That is, the smaller the temperature increase, the smaller the proportion of crops likely to be lost.

e) Mass migration

Changes in climatic conditions, including extreme weather events, can magnify migration within and between countries (IPCC, 2020). Such events may lead to food chain disruption, community displacement (e.g. due to flooding) and threatened livelihoods and negative economic impacts. It can also result in increased stressors for conflict. The impact of extreme weather events on economic outputs, including agriculture, has been found to increase the likelihood of conflicts and war, particularly in developing countries (Koubi, 2019). This can trigger mass migration resulting from rising sea levels and extreme weather events such as droughts (Koubi, 2019).

1.3 The built environment – from city scale, precinct to building scale

The built environment comprises various scales from city scale, which includes all built forms, and infrastructures above and can accommodate hundreds of thousands to tens of millions of people. The precinct or suburb scale is a smaller group of buildings accommodating a few thousand people, whilst the building scale may accommodate a few hundred people in the case of a high-rise residential building to a single person in a small house. Collectively, 40% of global GHG emissions are from the built environment, and therefore the built environment significantly contributes to climate change. Most building stock predates our awareness of climate change, which dates from 1987 and the Bruntland Report (Keeble, 1988). As such, it incorporates technologies from medieval to contemporary periods, materials that have been in use for centuries, to recently invented materials such as plastics. The stock includes low-technology buildings such as naturally ventilated low-rise buildings compatible with the local climate to provide comfortable accommodation to high-technology buildings utilising the latest AI and computer technologies to optimise user comfort and convenience. These high technologies can also optimise energy and water utilisation. Depending on scale and technology, there are possibilities to adopt sustainability and resilience measures at all three scales, from city to building level.

1.4 Transport, water and energy infrastructure

In order to function, the built environment and urban settlements are supported by transport, water and energy infrastructure. As settlements have grown in size and density, so too have the complexity and technological sophistication of these systems. These systems result in GHG emissions through fossil fuel petrol and oil consumption in cars, lorries, trains and planes. Water is consumed in sewerage and potable water supply to buildings and industrial production. Finally, the energy infrastructure is typically electricity generation based on fossil fuels, coal and gas. It does not need to be this way.

Transport includes road, rail and air transport, and for coastal and river-sited settlements, ferries and boats. All these forms of transport largely use fossil fuels for their power source, which results in substantial GHG emissions. Car ownership has grown significantly over the last few decades and is the norm in many countries and cities. Houses are developed on the basis of the need for car ownership to get to and from work, education, health, retail and leisure activities. Where compact settlements diminish the need for car transport, we have opted for sprawling urban development. Public transport systems, trains, buses and sometimes ferries allow for more people to be transported from place to place with typically lower GHG emissions per person per kilometre travelled. However, cities vary in their management and governance, with many places preferring private transport over financial and policy support for efficient, reliable and safe public systems. Other sustainable transport options include bicycle tracks to encourage people to travel safely and the adoption of electric vehicles, which requires local governments and councils to incorporate sufficient recharging stations for owners (Greene & Wegener, 1997; Zhao et al., 2020). Of course, electricity generation should be fossil free; solar and wind are suitable sources of power. There are forecasts for autonomous vehicles, which will be available via a phone app or similar, so users can travel from location to location without the need for car ownership and the real estate that comes with it, a garage. It is estimated that autonomous vehicles will be in use for up to 22 hours per day compared to private cars, which tend to be used less than 1 hour per day currently in most cities (Kim, 2018). If adopted en masse, autonomous vehicles could mean far fewer vehicles are needed than we currently have. This would extend the life of existing roads, tunnel and bridge infrastructure. In many cities, there is extensive disruption when existing transport infrastructure is upgraded to facilitate the transportation of greater numbers of vehicles. Extra lanes are added to existing motorways and roads, and new roads, tunnels and bridges are constructed to link new and existing suburbs. In terms of sustainability and resilience, built environments that offer a diverse range of public and private transport options are desired (Newton & Rogers, 2020). Local governments and councils are encouraged to adopt electric vehicle recharging infrastructure and to consider changes in planning that encourage less private vehicle ownership.

In 2020, GHG emissions fell an estimated 2.4 billion tons for the year, a 7% drop from 2019 and the largest recorded fall, which was triggered by global Covid-19 restrictions on travel and business-as-usual activities, according to research from the University of East Anglia, the University of Exeter and the Global Carbon Project (Le Quéré et al., 2020). Forty per cent of the fall is attributed to reductions in transport/travel alone. The researchers stated emissions were likely to rebound in 2021 if pre-2020 Covid-19 conditions returned and urged governments globally to prioritise transition to clean energy and policies that tackle climate change in their economic recovery plans. A decline in transportation activity drove the global drop in carbon emissions. The US had the largest drop in carbon emissions, 12%, followed by the European Union, 11%. India saw a drop in emissions of 9%, and China had a drop of 1.7% (Le Quéré et al., 2020). This is an excellent example of an acute shock to the resiliency of a system. Some supply chains to cities and urban settlements were disrupted as a result of the restrictions. The major impact was the reduction in business and holiday travel, which led to the positive impact of lower GHG emissions and the negative impact of job losses in some industries. It is a good example of the complex nature of sustainability and resilience, whereby both positive and negative impacts arise from events.

With water infrastructure, in some areas, lack of water is the issue, whereas, for others, excess water is a problem. Ageing water infrastructure can deteriorate over time, resulting in the loss of potable water through leaks. Clearly, effective maintenance and repair of existing systems are necessary to minimise loss. New developments should adopt the principles of Water Sustainable Urban Design (WSUD) – that is reducing flows and systems that deliver water. In buildings these include designs with rainwater harvesting for watering green spaces, lawns, greens roofs and walls. Internal designs include water recycling, whereby greywater (i.e. water from baths, sinks and showers) is reused to flush toilets and water gardens. It is possible to recycle water and to clean grey and blackwater (water from toilets) for reuse; however, currently, it is very energy-intensive, and if the energy source is fossil fuel based, it is not sustainable. This may change in future if non-fossil fuel renewable energy sources are used. In terms of resilience, if resilience is bouncing back from adverse events, the adoption of renewable energy ensures recycled water is not more harmful to the environment. For water-starved areas, rainwater retention and storage, the adoption of low water consumption technologies and reuse of water are all practical measures that ensure scarce resources last longer.

Energy infrastructure currently comprises electricity power stations powered by gas and coal, which results in large amounts of GHG emissions. The sustainable and resilient alternative is to use renewable energy in the form of wind and solar power. Depending on the location, some areas receive more sunlight or wind than other areas. This infrastructure can be located outside of the settlements or on buildings, or a combination of both. In order to make the energy supply more resilient, low-energy designs and retrofits to building envelopes in the form of insulation and reflective finishes, double or triple glazing specifications should be adopted to ensure the lowest possible demand on the sustainable renewable energy infrastructure (Ander, 2014; Bulut et al., 2021).

1.5 The response

This section of the chapter explores both the degree of response that is needed, if not demanded by our current circumstances with respect to climate change, and also the degree and quality of response to date.

1.5.1 The response that is needed

It is often argued that increasing regulation is the answer. Enacting high minimum standards levels the playing field with all stakeholders equally affected and ensures best practices are adopted. Others argue that a voluntary or market lead approach delivers greater innovation, as stakeholders compete to be the most sustainable. Trencher et al. (2016) explored innovative policy practices to advance building energy efficiency and retrofitting: approaches, impacts and challenges in ten C40 cities in the US and Asia-Pacific. They found mandatory and voluntary policy models with impacts and challenges and concluded innovation occurred without regulation or new policy invention and, by necessity, as generic models were adapted to local circumstances (Trencher et al., 2016). The sample revealed comprehensive regulation in Asia, experimentation with benchmarking in the US and voluntary approaches in Australia. Overall, environmental impacts emerged slowly and had attribution challenges. There was limited evidence of benchmarking programme effectiveness in reducing energy consumption in the short term but some indication of mid-term outcomes. The cap-and-trade model stood out by fostering large, sustained and attributable GHG emission reductions and retrofitting. Cap and trade is a term for a government regulatory programme designed to limit, or cap, the total level of emissions of chemicals, such as carbon dioxide. Proponents of cap and trade claim it is a viable alternative to a carbon tax. Trencher et al. (2016) found that the market and social impacts were highly significant, and there is a need to consider non-environmental impacts in policy evaluation. They conclude the complementary potential of voluntary and regulatory approaches to advancing energy efficiency and climate resilience needs to be explored further, as well as the potential for benchmarking programmes to transition to models mandating performance improvements, such as cap and trade (Trencher et al., 2016).

Another way to label this argument is 'the carrot or the stick' approach. Some argue that free markets will deliver greater results than mandated lower standards (Heffernan et al., 2021). In their analysis of policy pathways to an environmentally sustainable rental housing sector, Heffernan et al. (2021) reviewed international mandatory and voluntary approaches. Carrot policies included tax incentives, rebates and grants. Cusp policies, which were neither carrot nor stick, included loans, energy arrangements and improved rental rights, whereas the stick policies are minimum standards and mandatory disclosure (Heffernan et al., 2021). Minimum performance standards, rebates and tax incentives were found to be the most effective policy solutions identified, and the authors conclude policy mixes should include carrot and stick policies. Within the built environment, the sector, be it social or rental housing or commercial property, is influential with regard to whether a carrot or stick, a mandatory or voluntary or a hybrid policy is most effective. There is no one size fits all.

What is needed is collaboration. There is also a discussion to be had around needs and wants. The response to climate change and a stock of mostly inefficient ageing buildings, which predate minimum energy standards or sustainability, is largely overlooked. We need to act now whenever the opportunity arises to improve the sustainability, operational energy efficiency and resilience of our building stock. We may not want to do this. The question we need to ask ourselves is: for those owners who do not want to improve their existing buildings, how can we create circumstances to enable this to happen?

The November 2021 Paris COP26 ended with a global agreement to accelerate action on climate this decade, where COP agrees, for the first time, a position on phasing down unabated coal power (United Nations Climate Change Conference, 2021). Organisers claimed the COP26 summit brought parties together to accelerate action towards the goals of the *Paris* Agreement and the UN Framework Convention on Climate Change. Others, however, lament yet another lost opportunity to take effective action (The Guardian, 2021). They assert that limiting the 1.5 degree temperature increase is now no longer possible, and we are more likely to experience a plus 2 degree increase. There is evidence, however, that the longer we procrastinate and delay action, the more likely younger people will experience despair and anxiety about the future they will inherit.

1.5.2 The response to date: the scale of the problem for existing buildings

The existing stock of buildings contributes a substantial amount of GHG emissions and represents an excellent opportunity to make a significant reduction to GHG emissions if operational energy is reduced and materials with high embodied carbon are avoided. The opportunity to retrofit buildings occasionally arises during the life cycle, as the fabric or components of the building wear out to the point where replacement is necessary. Therefore there is a limit to what can be delivered through sustainable building retrofits if a business-as-usual approach is allowed to continue. To date, there is no mandatory requirement to undertake sustainable, energy-efficient building retrofits imposed on owners.

Interestingly, the impact of Covid-19 in 2020 and 2021 has required many people to work from home to prevent exposure to the virus whilst travelling to and from the workplace or being exposed in the workplace. There is now extensive debate about whether people will return to a five-day working week in a workplace (Savills, 2021). Many speculate that commercial buildings will be under-occupied and/or vacant. If this is the case, then some owners will seek to sell or convert or retrofit their stock to attract occupiers or buyers. They may adopt sustainable, resilient retrofits to increase their attractiveness to the market. There is a concern in the office sector that air-conditioned spaces may enable transmission of Covid-19 particulates from one area to another. The question arises: can we safely occupy spaces at the same density level as pre-Covid-19? Tests are being undertaken to look at different space plan layouts to ascertain pathways for air flow and particulates so that healthy building retrofits can be proposed. These two factors may create a market for increased building retrofit in the short-term future, which would create a unique opportunity to embrace low-carbon resilient, sustainable retrofits to mitigate the effects of climate change.

1.6 Summary of the challenges faced

This book is structured in three parts. Part 1 explores The Why – the challenge of climate change and the overarching need for extensive change. In Chapter 2, Sarah Sayce and Sara Wilkinson examine 'The philosophy and definition of retrofitting for resilience'. Retrofits are defined and distinguished from other forms of building alterations to differentiate a retrofit for resilience against a normal cyclical undertaking to bring a building back to extant occupational functional standard. The timing of retrofit is discussed, and 'deep' and 'light' retrofit approaches are defined. The chapter discusses what makes a 'resilient' building distinct from an 'environmentally friendly' or 'sustainable' one and the extent to which the terms are complementary. It argues that, if building resilience is to be achieved, sufficient to assist in meeting climate mitigation targets, not only is retrofitting buildings simply a necessity, but it is also a desirable social outcome, conserving as it may do the maintenance of the place, memory and culture. The chapter explores the extent to which retrofitting to preserve – or conserve – the social value of the building and its context is a philosophical, economical or legislative matter before highlighting the issues of the redevelop/retrofit decision. In conclusion, the chapter emphasises the paucity of the business-as-usual approach and outlines the radical alterations needed to our current conceptual understanding to deliver the necessary changes.

In Chapter 3, titled 'An inadequate building stock', Sara Wilkinson and Sarah Sayce describe the challenges facing the existing stock. The inadequacy most explored is the impact of age and associated energy inefficiency. This is a challenge that currently is unmet. However, retrofitting for low energy is only part of the picture: water conservation, flood protection and the ability to withstand fire are important criteria for resilient, sustainable buildings. As the health and well-being agenda gains traction and research connects chronic and fatal conditions with pollution and building defects, the quality of buildings is further brought into focus.

In 2020, a fatal and hitherto unacknowledged risk came to the fore: the transmission of disease within buildings through heating ventilation and air conditioning (HVAC) systems. Covid-19 showed transmission of the virus through HVAC systems was possible. This risk in some countries mandated office workers to operate from home to reduce exposure and transmission risk. Working from home worked well for some, and there are ongoing discussions about the viability of the return to work in offices post-pandemic and the benefits of allowing people to work from home. Even if a return to offices takes place, the type of space and the way it is operated will likely change, accelerating trends already identified (Harris, 2020). In the short term, this may lead to large office vacancies as leases terminate and leaseholders seek smaller demised spaces; in the longer term, buildings which cannot 'flex' to meet changing demand may become economically 'stranded'. Almost inevitably, there will be growth in 'inadequate stock'. Chapter 3 examines the mismatch of buildings to the social needs of communities: be that through the shift in transport means, such as the introduction of electric (or hydrogen) and, in time, autonomous vehicles, and the need to accommodate greater densities of population and to feed them locally. By outlining the drastic, radical changes needed, the chapter concludes that collectively this creates a multifaceted challenge to adapt and improve existing buildings.

The final chapter in Part 1 is titled 'Understanding Vacancy in the office stock'. Dr Gillian Armstrong argues that given the uncertain impacts of Covid-19, the need for greater vacancy understanding is particularly urgent. Current predictions are for high levels of commercial building vacancy and an increased risk of premature obsolescence. This chapter makes a case for a more nuanced understanding of vacancy as an evidence base for mitigating obsolescence and building urban resilience. An analysis of vacancy data challenges the accepted wisdom of building urban resilience by converting vacant office buildings to new uses. The chapter offers suggestions on how to advance vacancy knowledge and describes a tool for policymakers to quantify vacancy, known as Vacancy Visual Analytic Method (VVAM) (Armstrong et al., 2021). Finally, this chapter highlights the usefulness of vacancy as an essential evaluation tool in policy development to address chronic stresses and acute shocks experienced by cities.

Part 2 explores The What – setting out what is needed to deliver the changes identified in the why part of the text. Jeroen van der Heijden sets the scene in Chapter 5 titled, 'A governance response: from coercion to persuasion to embracing diversity?' van der Heijden claims that seeking to achieve retrofits, government responses can and have been coercive, persuasive or both, and range from punitive tax regimes and statutory requirements to 'nudge' techniques and voluntary programmes. This chapter analyses a range of measures in different jurisdictions and across the spectrum of interventions (such as taxes, certification requirements, statutory obligations and economic incentives). It assesses whether such measures are sufficient in light of the great urgency of climate change and awareness of health and well-being in the light of Covid-19. It argues that not enough is being done to shift the pendulum from coercive to persuasive techniques and suggests ways in which governments should seek higher levels of effectiveness through an overhaul of the building regulatory system. This overhaul would involve combining coercive and persuasive interventions and targeting different groups of property owners and users with tailored regulatory and governance interventions.

Chapter 6, 'Financing Retrofits', written by Zsolt Toth, Ursula Hartenbeger and Sarah Sayce, examines the ways financing building retrofits is evolving to assist owner-occupiers and investment owners to gain finance from either public or private financiers, or both. It concludes the situation is changing rapidly, and arguments previously made that funding was unavailable have started to be overcome. However, it concludes that more needs to be done and posits some recommendations to aid both the speed and quantum of progress.

In Chapter 7, 'Technological Solutions', Sara Wilkinson and Sam Organ explore innovations in technology that could bring significant change to the carbon footprint of existing buildings to address the critical issues set out in Part 1. Extreme times demand consideration of new approaches, such as the end of fossil fuels as primary energy sources. Wilkinson and Organ argue this is already happening through a combination of investments (public and private) and the impact of legislation and policy. In evidence, they refer to the UK's national electricity grid experiencing 18 days of fossil-free energy generation in April 2020 (The Guardian, 2020). They ask, 'can buildings derive sufficient energy from non-fossil sources, and will it need rapidly accelerated rates of retrofitting to achieve the change from fossil fuels and a reduced carbon footprint?' Other benefits such as bioremediation of greywater in buildings are described. Innovations in sensors, technology, artificial intelligence and robots, they claim, may facilitate greater adoption of green infrastructure, which, if adopted 'en masse' in city centres, mitigates the urban heat island effect. One method of distinguishing innovations is whether they are considered low or high tech: that is, whether they have little or no reliance on computerised technology (low tech); or whether features such as computer technologies, sensors or artificial intelligence (AI) are integrated into a building ('high tech') for performance optimisation. Drawing on innovations from a range of disciplines, this chapter sets out new technologies, new ideas and new ways of retrofitting existing buildings to deliver more sustainable and resilient outcomes.

In Chapter 8, Hilde Remøy and Sara Wilkinson focus on repurposing and adaptation. Social and technological change will always affect buildings and how we design and use them. An example is the advent of driverless vehicles and a sharing economy model, which is predicted to decrease car ownership. Currently, most cars globally are in use less than 1 hour per day, so we build structures to park them for 95% of the day. As this decline occurs, large amounts of car parking space will become redundant. Elsewhere the retail sector and the high street are suffering from the advent of online shopping. So, what can be done to repurpose and adapt this stock? Remøy and Wilkinson argue some changes are unpredictable and fast, known as 'acute shocks' in resilience parlance. Others are slow and ongoing or chronic. In 2020 the globe experienced a health shock in the form of Covid, which quickly turned into a global pandemic, a health crisis that spread rapidly and was exacerbated by international air travel. As a result, global travel shut down to essential travel only, with people required to quarantine on arrival in many countries. Soon economic impacts were felt; people were told to stay home and work and not to socialise outside the home. Retail switch to online shopping accelerated to minimise exposure to the virus, and restaurants turned to home delivery models. Socialising at sporting, cultural, music and arts events ceased. In 2021 this is impacting our existing building stock, and the full outcome is yet to be realised; however, we can see from previous changes what can happen and, in this way, explore what might happen in the future. This chapter examines innovative ideas for repurposing and adapting redundant stock for new uses which meet revised needs and demands. For example, urban food production, shared affordable and alternative housing are some options explored.

The last chapter in Part 2 is Chapter 9, 'Heritage: learning from and preserving the past'. Here Sara Wilkinson and Shabnam Yazdani Mehr explain why heritage buildings are important to remind us of our history, and the need to conserve and preserve remains important. There are many ways in which these buildings physically embody resilience, and they can be retrofitted and adapted sustainably. Many heritage buildings adopted what we now consider sustainable materials and technologies as they predate industrialised methods of production and the reliance on high levels of energy and mechanisation for operation, and therefore, there is much to be learned from them. Over time some buildings are adapted and retrofitted within the use, whereas others are converted or undergo adaptive reuse. At this point, issues around the place and location, or 'genius loci' and authenticity, become important. These changes are a result of the prevailing legal, technological, social, economic and environmental drivers prevailing in the location at that point in time. This chapter explores what we can learn from heritage stock to make their retrofit resilient; and what we can learn and transfer into the retrofit of other, non-heritage stock. A model for assessing adaptive reuse of heritage buildings and a checklist for identifying and preserving 'genius loci' in adaptive reuse are proposed.

Finally, Part 3, Chapter 10, concludes the text and sets out a manifesto for change. Building on the preceding chapters, this last chapter presents a manifesto of recommendations for policymakers, educationalists, professional bodies and practitioners. Whilst it may be speculative in some ways, the intent is to underscore the conviction that a business-as-usual model can no longer work; it will argue that the responses to date to dealing with building adaptation are too timid – and that this lack of real commitment and drive could be argued, to quote Greta Thunberg, is 'beyond absurd'.

1.7 Conclusion

This first chapter has set the scene in respect of global climate change causes and the social, environmental and economic impacts and the considerable contribution of the built environment to global warming. It has highlighted the need for more sustainable retrofits to mitigate these impacts and to contribute to lessening waste, reducing building-related water and energy consumption, to adopting a circular economy approach. The rationale for the three parts to the book structure was explained. Part 1 explores The Why – the challenge of climate change and the overarching need for extensive change. Part 2 explores The What – setting out what is needed to deliver the changes identified in the why part of the text, and finally, Part 3 concludes the text and sets out a manifesto for change.

References

Ander, G. D., 2014. Windows and glazing. *Whole Building Design Guide*. Available from https://www.wbdg.org/resources/windowsand-glazing.

Armstrong, G., Soebarto, V. and Zuo, J., 2021. Vacancy visual analytics method: Evaluating adaptive reuse as an urban regeneration strategy through understanding vacancy. *Cities*, *115*, p. 103220.

Ayanlade, A., Radeny, M. and Morton, J. F., 2017. Comparing smallholder farmers' perceptions of climate change with meteological data: A case study from southwestern Nigeria. *Weather and Climate Extremes*, *15*, pp. 24–33.

Bernstein, L., Bosch, P., Canziani, O., Chen, Z., Christ, R. and Riahi, K., 2008. *IPCC, 2007: Climate Change 2007: Synthesis Report*. IPCC, Geneva.

Bill Gates and Melinda French Gates Foundation, 2020. 2020 Goalkeepers report. Covid-19 – A global perspective. Available from www.gatesfoundation.org/goalkeepers/report/2020-report/#CollaborativeResponse. Accessed 20 December 2021.

Bulut, M., Wilkinson, S., Khan, A., Jin, X. H. and Lee, C. L., 2021. Thermal performance of retrofitted secondary-glazed windows in residential buildings – Two cases from Australia. *Smart and Sustainable Built Environment*. SASBE 03 2021 0050. https://doi.org/10.1108/SASBE-03-2021-0050.

Clayton, J., Devaney, S., Sayce, S. and van de Wetering, J., 2021. *Climate Risk and Commercial Property Values: A Review and Analysis of the Literature*. UNEP FI. Available from unepfi.org/publications/investment-publications/climate-risk-and-commercial-prop- erty-values/.

Clayton, S., Devine-Wright, P., Stern, P. C., Whitmarsh, L., Carrico, A., Steg, L., Swim, J. and Bonnes, M., 2015. Psychological research and global climate change. *Nature Climate Change*, *5*(7), pp. 640–646.

Climate Emergency Declaration, 2020. Available from https://climateemergencydeclaration.org/climate-emergency-declarations-cover-15-million-citizens/. Accessed 30 December 2020.

Eriksen, S., Schipper, E. L. F., Scoville-Simonds, M., Vincent, K., Adam, H. N., Books, N., Harding, B., Khatri, D., Lenaerts, L., Liverman, D., Mills-Novoa, M., Mosberg, M., Movik, S., Muok, B., Nightingale, A., Ojha, H., Sygna, L., Taylor, M., Vogel, C. and West, J. J., 2021. Adaptation interventions and their effect on vulnerability in developing countries: Help, hindrance or irrelevance? *World Development*, *141*(105383).

Geissdoerfer, M., Savaget, P., Bocken, N. M. and Hultink, E. J., 2017. The circular economy – A new sustainability paradigm? *Journal of Cleaner Production*, *143*, pp. 757–768.

Greene, D. L. and Wegener, M., 1997. Sustainable transport. *Journal of Transport Geography*, *5*(3), pp. 177–190.

The Guardian, 22 November 2020. Britain breaks record for coal-free power generation. Available from https://www.theguardian.com/business/2020/apr/28/britain-breaks-record-for-coal-free-power-generation.

The Guardian, 11 November 2021. Cop26 targets too weak to stop disaster, say Paris agreement architects. Available from www.theguardian.com/environment/2021/nov/11/cop26-targets-too-weak-to-stop-disaster-say-paris-agreement-architects. Accessed 16 December 2021.

Guntukula, R., 2019. Assessing the impact of climate change on Indian agriculture. Evidence from major crop yields. *Journal of Public Affairs*, *20*(1), pp. 1–7.

Harris, R., 2021. *London's Global Office Economy: From Clerical Factory to Digital Hub*. Routledge, London.

Heffernan, T. W., Daly, M., Heffernan, E. E. and Reynolds, N., 2021. The carrot and the stick: Policy pathways to an environmentally sustainable rental housing sector. *Energy Policy*, 148, p. 111939.

Ide, T., Fröhlich, C. and Donges, J. F., 2020. The economic, political, and social implications of environmental crises. *American Meteological Society*, 101(3), E364–E367.

IPBES, 2019. Global assessment report on biodiversity and ecosystem services. Available from https://ipbes.net/global-assessment. Accessed 7 December 2021.

IPPC, 2007. *Impacts, adaptation and vulnerability: Contribution of Working Group II to the Fourth Assessment Report of the Intergovernmental Panel on Climate Change* (Vol. 4). Cambridge University Press, Cambridge.

IPCC, 2020. Climate change and land – summary for policymakers. Available from www.ipcc.ch/site/assets/uploads/sites/4/2020/02/SPM_Updated-Jan20.pdf. Accessed 16 December 2021.

Islam, N. and Winkel, J., 2017. Climate change and social inequality. *UN Department of Economic and Social Affairs (DESA) Working Papers*, No. 152, pp. 1–32.

Keeble, B. R., 1988. The Brundtland report: 'Our common future'. *Medicine and War*, 4(1), pp. 17–25.

Kenter, J. O. and Dannevig, H., 2020. Adaptation to climate change-related ocean acidification: An adaptive governance approach. *Ocean and Coastal Management*, 191(105176).

Kim, T. J., 2018. Automated autonomous vehicles: Prospects and impacts on society. *Journal of Transportation Technologies*, 8(3), p. 137.

Kirchherr, J., Reike, D. and Hekkert, M., 2017. Conceptualizing the circular economy: An analysis of 114 definitions. *Resources, Conservation and Recycling*, 127, pp. 221–232.

Koubi, V., 2019. Climate change and conflict. *Annual Review of Political Science*, 22, pp. 343–360.

Le Quéré, C., Jackson, R. B., Jones, M. W., Smith, A. J., Abernethy, S., Andrew, R. M., De-Gol, A. J., Willis, D. R., Shan, Y., Canadell, J. G. and Friedlingstein, P., 2020. Temporary reduction in daily global CO_2 emissions during the COVID-19 forced confinement. *Nature Climate Change*, pp. 1–7.

Lyster, R., 2017. Climate justice, adaptation and the Paris agreement: A recipe for disasters? *Environmental Politics*, 26(3), pp. 438–458.

Mase, A. S., Cho, H. and Prokopy, L. S., 2015. Enhancing the social amplification of risk framework (SARF) by exploring trust, the availability heuristic, and agricultural advisors' belief in climate change. *Journal of Environmental Psychology*, 41, pp. 166–176.

Newton, P. W. and Rogers, B. C., 2020. Transforming built environments: Towards carbon neutral and blue-green cities. *Sustainability*, 12(11), p. 4745.

Ord, T., 2020. *The Precipice: Existential Risk and the Future of Humanity*. Hachette Books, New York.

Rogelj, J., Huppmann, D., Krey, V., Riahi, K., Clarke, L., Gidden, M., Nicholls, Z. and Meinshausen, M., 2019. A new scenario logic for the Paris Agreement long-term temperature goal. *Nature*, 573(7774), pp. 357–363.

Rogelj, J., Shindell, D., Jiang, K., Fifita, S., Forster, P., Ginzburg, V., Handa, C., Kheshgi, H., Kobayashi, S., Kriegler, E. and Mundaca, L., 2018. Mitigation pathways compatible with 1.5 C in the context of sustainable development. In *Global Warming of 1.5 C* (pp. 93–174). Intergovernmental Panel on Climate Change, Geneva.

Savills, 2021. *Impacts. The Future of Global Real Estate*, Issue 4, 2021. Available from https://www.savills.com/impacts/index.html?__hstc=122678100.44a1f81448a0077aa7b93e9ac

60ba1b2.1616323131629.1639373389528.1639465792208.96&__hssc=122678100.8.1
639465792208&__hsfp=1486769359.

Shukla, P. R., Skea, J., Calvo Buendia, E., Masson-Delmotte, V., Pörtner, H. O., Roberts, D. C., Zhai, P., Slade, R., Connors, S., Van Diemen, R. and Ferrat, M., 2019. *IPCC, 2019: Climate Change and Land: An IPCC Special Report on Climate Change, Desertification, Land Degradation, Sustainable Land Management, Food Security, and Greenhouse Gas Fluxes in Terrestrial Ecosystems.* IPCC, Geneva.

Stern, N. and Stern, N. H., 2007. *The Economics of Climate Change: The Stern Review.* Cambridge University Press, Cambridge.

Tol, R. S., 2002. Estimates of the damage costs of climate change. Part 1: Benchmark estimates. *Environmental and Resource Economics, 21*(1), pp. 47–73.

Trencher, G., Broto, V. C., Takagi, T., Sprigings, Z., Nishida, Y. and Yarime, M., 2016. Innovative policy practices to advance building energy efficiency and retrofitting: Approaches, impacts and challenges in ten C40 cities. *Environmental Science & Policy, 66*, pp. 353–365.

UN Action, 2019. Species extinction not just a curiosity: Our food security and health are at stake. Available from www.unep.org/news-and-stories/story/species-extinction-not-just-curiosity-our-food-security-and-health-are-stake. Accessed 20 December 2021.

UN Climate Change Conference, 2021. COP26 keeps 1.5C alive and finalises Paris agreement Available from https://ukcop26.org/cop26-keeps-1-5c-alive-and-finalises-paris-agreement/. Accessed 16 December 2021.

UNFCCC, 2020. The Paris agreement. Available from https://unfccc.int/process-and-meetings/the-paris-agreement/the-paris-agreement. Accessed 30 December 2020.

Wheeler, R. and Lobley, M., 2021. Managing extreme weather and climate change in UK agriculture: Impacts, attitudes and action among farmers and stakeholders. *Climate Risk Management, 32*(2021).

Whyte, K., 2019. Too late for indigenous climate justice: Ecological and relational tipping points. *WIREs Climate Change, 11*(1).

Wilby, R. L., 2007. A review of climate change impacts on the built environment. *Built Environment, 33*(1), pp. 31–45.

World Metrological Organization, 2021. WMO recognizes new arctic temperature record of 38°C. Available from https://public.wmo.int/en/media/press-release/wmo-recognizes-new-arctic-temperature-record-of-38°c. Accessed 20 December 2021.

Zhao, X., Ke, Y., Zuo, J., Xiong, W. and Wu, P., 2020. Evaluation of sustainable transport research in 2000–2019. *Journal of Cleaner Production, 256*, p. 120404.

2 The philosophy and definition of retrofitting for resilience

Sara Wilkinson and Sarah Sayce

2.1 Introduction

This chapter starts by defining retrofits and distinguishing this from other forms of building alterations in order to distinguish a retrofit for resilience against a normal cyclical undertaking to bring a building back to an extant occupational functional standard. It considers the timing at which retrofitting takes place and distinguishes 'deep' and 'light' retrofit approaches. The chapter then expands into a discussion as to what constitutes a 'resilient' building as distinct from an 'environmentally' friendly' or 'sustainable' one and the extent to which the terms are complementary. It argues that if building resilience is to be achieved, sufficient to assist in meeting climate mitigation targets, not only is retrofitting buildings simply a necessity, but it is also a desirable social outcome, conserving as it may do the maintenance of place, memory and culture.

The chapter debates what happens when place and community are lost, for example, as in the London Docklands, and provides some examples of where buildings have achieved a new life. The chapter explores the extent to which retrofitting to preserve – or conserve – the social value of the building and its context is a philosophical, economical or legislative matter before highlighting the issues of the redevelop/retrofit decision. Finally, the chapter emphasises where the business-as-usual approach is lacking and outlines the radical and drastic alterations needed to our current conceptual understanding to deliver the necessary changes.

2.2 Types of retrofitting

2.2.1 Retrofit

Retrofitting is the process of modifying something after it has been manufactured: in this case, a building. There are various degrees of retrofit, from light to extensive or deep retrofit. Deep retrofit is a term associated with extensive energy upgrades to the building fabric and envelope (Rocky Mountain Institute, 2020).

Retrofitting a building involves changing its systems or structure after its initial construction and occupation (City of Melbourne, 2021). This work can not only restore functionality but also improve both the amenities of the building

DOI: 10.1201/9781003023975-3

for the benefit of its occupants and the technical performance environmentally. As technology develops over time, building retrofits can significantly reduce energy and water usage and improve sustainability performance; further, like climate change, for example, rising temperatures, retrofitting may be essential if the building is to remain functional. This has been brought sharply into focus, as weather patterns start to change more rapidly than previously predicted, making some buildings constructed only a few years ago already requiring technology upgrades to combat extreme heat.[1] There are several terms used in different countries to describe retrofit; these are defined later. What is relevant to the discussion in this book is the notable absence of a connection between definitions of retrofit and the principles of circularity. This point is explored later in the chapter.

2.2.2 Refurbishment

In the UK, the term 'refurbishment' is used to describe building adaptation and retrofit (Wilkinson et al., 2014). During a refurbishment, a building is improved above and beyond its initial condition or brought up to extant building compliance standards. This is a good example of different terms being used to describe the same set of activities. According to the City of Melbourne (2021), refurbishments are often focused on aesthetics and tenant amenities but can also include upgrades to a building's mechanical and electrical systems to improve energy and water efficiency. Therefore, it can be argued that the terms are sometimes useable interchangeably – but it is always critical to understand what is being described to ensure clarity.

2.2.3 Renovation

Similarly, renovations are very similar to refurbishments, and the terms are sometimes used interchangeably. The major difference is the term 'renovation' applies specifically to buildings, while 'refurbishment' does not (City of Melbourne, 2021). As with refurbishments, renovations often focus on aesthetics and tenant amenities but may include upgrades to mechanical and electrical systems and, therefore, potentially have a positive effect on energy and water efficiency (Gustafsson et al., 2017). Renovation, as defined in the *Oxford Dictionary*,[2] also applies to the process of bringing buildings that have fallen into disrepair up to the expected 'norms' of performance and specification; it implies repair, bringing into condition and some rebuilding.

2.2.4 Retro-commission

Commissioning is the process whereby newly installed building services are tested and adjusted to ensure they are functioning correctly prior to formal handover from the project team. Retro-commissioning is performing the same process on a building that has been operational and occupied for a period with the purpose of ensuring it continues to meet the design intent and the needs of the occupants. Some experts recommend retro-commissioning or re-commissioning a building once every three to five years. This is considered vital given that many new builds, and retrofitted buildings, in practice, fail to operate at designed standards: the so-called energy performance gap. For a review of the causes of the gap, see Van Dronkelaar et al. (2016).

If a building is not properly commissioned, has had changes to its systems and operating conditions made since commissioning or has had its performance degrade over time, retro-commissioning may make its existing systems more efficient (Jump et al., 2007). Further, as the ability to monitor water and energy use improves using 'smart' meters and other devices aimed at enhancing efficiency and reducing resource use, retro-commissioning makes increasing environmental – and economic – sense.

As with initial commissioning, retro-commissioning can be carried out by the contractors who installed the building's mechanical systems or by a third party who specialises in commissioning.

2.2.5 Tune-up

Finally, 'building tune-up' is a generic term that may encompass maintenance on the building's existing systems or aspects of retrofitting and retro-commissioning. Many cities globally have implemented building audits to inform 'tune-up' programmes with the goal of improving energy and water use efficiency (Ballinger, 2020); however, although a study of their adoption in New York (Kontokosta et al., 2020) showed limited success in reducing energy use, whether such schemes work may be considered questionable.

2.3 Defining a resilient building

Defining a resilient building is challenging; it is also a comparatively new addition to the real estate lexicon. It is critical to the arguments put forward in this book that it is understood to be distinct from, for example, a 'green' or environmentally friendly building. To many, a sustainable building is viewed only in terms of energy efficiency and/or water saving technologies; others focus on the importance of materials and embodied carbon in reducing the whole life footprint of the building. Further definitions of sustainable buildings extend beyond the fabric to user behaviours which exercise considerable, variable impacts on the amount of delivered building performance. Whilst some aspects of 'sustainability' can be built in through design and original construction, the process of retrofit is key to climate targets, given that most buildings were not built with carbon neutrality in mind and are certainly not 'future-proofed'. Indeed, it is argued that in countries, such as the UK, where old stock dominates, up to 85% of buildings require extensive work to make them resilient in carbon terms. Some, of course, for physical or heritage reasons, simply cannot be made truly carbon neutral.

However, the concept of resilience in buildings is altogether wider than that of zero carbon in use. It places the building into its physical, economic and social contexts; however, it is argued to be distinct from a 'green building' (Hewitt et al., 2019). Even sustainability could still be as contested today as it was nearly 30 years ago, when Cook and Golton (1994) argued that definitional disputes could not be settled by appeal to more reference to empirical evidence, linguistic usage or the canons of logic alone. Sustainability is also based on our lived experience and beliefs, with people interpreting information differently (Wilkinson, 2012). When applied to buildings, sustainability embraces longevity, adaptability, low resource impact, location sensitivity and the notion of likeability – or appeal to our sensitivities.

Environmental impacts can result from products specified, how the building is constructed and how it is operated during the life cycle. The life cycle also includes maintenance, repairs and retrofits (or refurbishments) and, finally, the end-of-life demolition and potentially reusing or recycling and disposal of the building materials, fabric and structure. This is summarised in Figure 2.1. Phillips et al. (2017) contend that each stage has sustainability issues and impacts and resilience issues and impacts.

Figure 2.1 considers sustainability in the context of an evaluation of resilient strategies. Stages A1–B6 form the basis of the sustainability evaluations. Greyed processes (B7–C4) are those which move from a linear to a circular life cycle but which were not considered by Phillips et al. (2017).

In the disciplines of engineering and construction, resilience is the ability to absorb or avoid damage without suffering complete failure and is an objective of design, maintenance and restoration for buildings and infrastructure as well as communities. Moazami et al. (2019) posited two definitions for robust and resilient building in respect of dealing with and preparing for climate uncertainty:

Definition 1: A robust building is a building that, while in operation, can provide its performance requirements with a minimum variation in a continuously changing environment (Moazami et al. (2019).

Robustness emerged in the 1940s as a concept that meant products, technologies or products had the characteristic whereby they were not sensitive to factors causing variability and ageing. Significantly, this period coincided with the

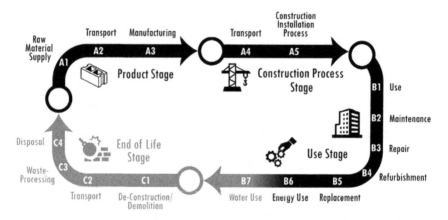

Figure 2.1 Building lifecycle stages from Phillips et al. (2017:299), who adapted it from BS EN 15978

Second World War and/or the period immediately after when resources were scarce globally.

Definition 2: A resilient building is a building that not only is robust but also can fulfil its functional requirements (withstand) during a major disruption. Its performance might even be disrupted but has to recover to an acceptable level in a timely manner in order to avoid disaster impacts.

Two critical variables were identified in their research, functional and performance requirements. The *functional requirements* define what a building has to do, and the *performance requirements* determine how well a functional requirement has to be done (de Wilde, 2019).

Using these definitions, Moazami et al. (2019) proposed a conceptual figure to illustrate the relationship of robustness and resilience in respect of functionality and performance. Figure 2.2 shows that robustness is part of the typical expectation of buildings which are weakened by unforeseen and extreme events but which may recover to an acceptable or even design performance level. Resilience, on the other hand, covers the whole scope of foreseen and unforeseen events.

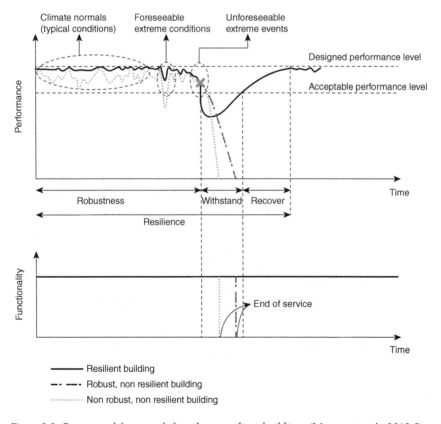

Figure 2.2 Conceptual framework for robust, resilient buildings (Moazami et al., 2019:5)

However, this framework covers only climate-related events, such as flooding and extreme heat, whereas true resilience should include not only all climate hazards, including storms, but also economic and social impacts. Economic impacts, such as variable demand, tend to lead to the need to repurpose, whereas social impacts, including legislative and technology change, can require changes to meet the needs to protect human health, such as witnessed in the Covid-19 crisis, as well as legislation to ensure fair access to buildings by those with disability. These factors combine to make the process of, and design for, retrofitting critical if our buildings are to 'live' for their maximum timespan – as Brand argued so many years ago (Brand, 1997). We need to help our buildings learn – and then we need to learn with them and respect them, seeking to prolong their existence where appropriate to aid maintenance of cultural history. This theme is picked up later in Section 2.6.

At this point, before moving on, it is important to distinguish between in-use adaptation and across the use. The latter – for example, retrofitting an office to a flat – is an example of across use adaptation but configuring a standard office into one that supports, for example, co-working is in-use adaptation. Where this can be achieved, it is more likely to retain the social identity of an area.

It is useful before finalising this debate to make mention of an earlier conceptualisation of resilient buildings that were put forward by ARUP as part of the 100 Resilient Cities Agenda,³ which are set out in Table 2.1.

In conclusion, in seeking to distinguish resilient buildings from sustainable ones, it is useful to refer to the definition given by Jennings et al. (2013) (quoted in Phillips et al., 2017:296). This concludes that a resilient building can '*resist physical damage, may be quickly and cost-effectively repaired if damaged, and maintains key building functionality either throughout a disruptive event or restores a target operation level more quickly after such an event occurs*'.

Taking this framework and the literature detailed earlier, but also drawing from wider literature, including the economic context explored by Ellison and Sayce (2007), it is posited that truly resilient buildings are those which are

1. at low risk from external hazards and physical climate events such as wildfires, drought or flooding or sea-level rise;
2. designed or adapted to use renewable energy and natural heating/cooling systems such that they can operate with minimal or no support from external sources and preferably with 'back-up supply';
3. durable in structure but adaptable both in-use and across uses;
4. conducive to good health and well-being, and located away from unhealthy external, polluting environments;
5. accessible and allow easy connectivity to the markets or communities they serve, and with low reliance on solely road access; and
6. able to recover quickly from external shock, that is, are able to be repaired using readily available local materials and labour.

Table 2.1 Qualities of resilient buildings

Quality	Characteristics
Reflective	where buildings are able to accommodate uncertainty and change, with the ability to evolve based on emerging evidence.
Robust	where buildings are well conceived, built and managed so they can withstand impacts without significant damage and loss of function. Avoiding over-reliance on any single component makes the building less vulnerable to catastrophic collapse.
Redundant	where buildings have the capacity to accommodate disruption and/or demand surges. Diversity increases capacity and ways of achieving different functions.
Resourceful	this is the ability of people and organisations owning and managing buildings to find alternate ways to meet needs and achieve goals when experiencing shocks or stresses. Resourcefulness is vital to restore the building functionality of critical systems when severe conditions prevail.
Inclusive	where the need for broad consultation and engagement of all members of society is recognised. Intra-societal equity is a fundamental component of resilience. Adoption of inclusion results in shared ownership and joint vision to build resilience.
Flexible	acknowledging change is inevitable, building design should accommodate changes to technology and space plans.
Integrated	alignment and integration for consistency in decision-making across all scales. Here, scales include the design team, engineers, planners, contractors, as well as building users and regulatory bodies (planning and building compliance). Effective knowledge exchange between stakeholders enables them to function collectively and respond rapidly through effective communications.

(Source: Authors Rockefeller Foundation 2022).

From this, it can be argued that resilience is a moving target that requires continued investment; this is explored later.

2.4 The case for retrofitting: overcoming obsolescence

There is nothing new in the requirement for retrofitting: *Panta Rhei*, or 'life is flux', is attributed to Heraclitus of Ephesus, a Greek philosopher who made this profound statement around 500 years BC (Ancient History, 2021). Some 2,500 years later, this remains true: the only constant is change. Change, as we know, can be slow or fast, predictable or unpredictable. Furthermore, change has many dimensions: it can be political, economic, social, technological, legal and/or environmental. Change can have single or multiple causes and consequences.

It has long been recognised that, over time, buildings may experience a decline in utility or usefulness (Baum, 1993; Mansfield & Pinder, 2008; Pinder & Wilkinson, 2001; Salway, 1986;), and this is recognised as a sustainability issue (Reed & Warren-Myers, 2010). It used to be thought that the lifespan of a building would be determined by the longevity of its fabric and that problems of obsolescence were relatively innocuous. Bowie's short seminal article (Bowie, 1982) demonstrated the fallaciousness of that argument. Today, most building types are increasingly prone to obsolescence because of the functional, economic and social requirements being placed on them by economic shifts, revolutionary technologies and emerging cultures (Nanyakkara et al., 2021).

From the point of first occupation, buildings physically deteriorate, and the capital invested in them undergoes a gradual process of devaluation; as buildings age and decay, they suffer diminished utility, requiring a constant stream of capital investment (Mansfield & Pinder, 2008; Bryson, 1997). Nevertheless, physical deterioration of buildings is largely a function of time and use and can be controlled to some extent by selecting appropriate components and materials at the design stage and by correct, planned maintenance (Chanter & Swallow, 1996; Thomsen & Van der Flier, 2011). Further, a building that retains economic or social relevance, possibly through locational advantage or historical importance, will present a case for investment to combat physical deterioration, as the existence of many heritage buildings testifies. Furthermore, lifecycle cost analysis (LCA) facilitates choice between alternative design options (Kishk & Al-Hajj, 200099) and helps identification of issues that, potentially, lead to deterioration (Crawford, 2011).

Physical deterioration should not be confused with a building's decline in utility due to a failure to satisfy new needs created by changes in equipment, materials, style, laws and the many other forces that cause a building to lose desirability in the eyes of its user and hence suffers from obsolescence (Grover & Grover, 2015; Mansfield & Pinder, 2008; Pourebrahimi et al. (2020); RICS, 2018; Sayce et al., 2004). More specifically, obsolescence describes a relative decline in the utility of a building that does not result directly from physical usage, the action of the elements or the passage of time (Baum, 1993). Instead, obsolescence is caused by changes in peoples' needs and expectations regarding the use of a particular building (Thomsen & Van der Flier, 2011). Utility – the sense of usefulness, desirability or satisfaction – is central to the concept of obsolescence; if a building does not provide utility, it will be considered obsolete (Smith et al., 1998). The Golf Club House, Rochford Hall, in Rochford, Essex, England, is a good example of an 800-year-old building that has had several uses during its life cycle, which inevitably became obsolete (Wikipedia, 2021). These uses range from a medieval hall, to a fortified Tudor house and Boleyn family home, complete with moat, to its current use as a golf club.

However, there is no objective measure of utility for buildings, and, if there was, it is unlikely that the changes over time would be represented by a straight line; the pattern of change would be more complex (Mansfield & Pinder, 2008; Khalid, 1993). The lack of an objective measure of building utility presents two problems. The first problem is that obsolescence is difficult to control. It was identified many years ago by Ashworth (1997) that in contrast to the gradual process of physical deterioration, obsolescence occurs at irregular and unpredictable intervals and is concerned with uncertain events, such as changes in fashion and technology, as well as innovation in the design and use of buildings. The range of variables and the unpredictability of some of these influences imply that a general model of obsolescence is not feasible (Thomsen & Van der Flier, 2011; Golton, 1989), and the scope for preventative action appears limited (Salway, 1986). Despite this, there have been recent attempts to find systematic ways to assess the rate of depreciation in value (Chen et al., 2017). The second problem is that obsolescence is a relative matter, which means that rational, consistent measures are very difficult to produce and are subjective (Thomsen & Van der Flier, 2011). This subjectivity derives from the fact that perceptions of obsolescence change relative to a particular situation or condition and vary according to the viewpoint or interest of the observer; obsolescence is a function of human decision rather than a consequence of 'natural' forces (Mansfield & Pinder, 2008; Cowan, 1970).

Historically, the problem of measurement has been overcome by focusing upon the financial impact of obsolescence by measuring obsolescence in terms of a real or nominal decrease in building value. Baum (1993) and Khalid (1993) used the financial impact of obsolescence to measure the effects of obsolescence on the depreciation of office buildings in the investment property market. The limitation of the financial approach allows us to isolate two forms of obsolescence.

Building obsolescence 'occurs when a building's stream of rental payments bears little relationship to the rental payments usually obtained from that location' (Bryson, 1997:1446). It is therefore concerned with buildings' physical characteristics, as determined by design and specification. Locational obsolescence occurs when buildings located within a particular area suffer from devaluation because the area is seen as less attractive by current or prospective occupiers (Bryson, 1997). Locational obsolescence results from changing expectations of infrastructure, communications and environmental conditions (Cowan, 1970; Lichfield et al., 1968). It is much more difficult for an individual building owner or user to remedy the causes of locational obsolescence, whereas building obsolescence can often be remedied by retrofit (Wilkinson & Remoy, 2018; Wilkinson et al., 2014) or imaginative reuse and repurposing.

Renewal, which involves demolition and replacement, is time-consuming and expensive. Furthermore, renewal under current practices typically results in most materials and components going to landfill waste. This is inherently unsustainable; the aim should be to promote circularity where possible. This is the argument increasingly put forward – notably recently by the Royal Institute of British Architects (RIBA), who argue that there should be a presumption in favour of retrofit as 'every year 50,000 buildings are demolished in the UK, producing 126 million tonnes of waste, which represent two-thirds of the UK's total waste'.[4] Retrofit retains the embodied carbon in building materials and typically costs much less than a renewal and can be delivered in shorter time frames. Increasingly, in the debate around retrofit terms such as resilient and sustainable are incorporated (Wilkinson & Remoy, 2018). What do these terms mean, and moreover, do they compete, or are they compatible and overlapping?

Since many of the aforementioned reports were written and debates rehearsed, both climate change and Covid-19 have become major relevant issues. They provide examples of change that require immediate and urgent responses; the former is multi-cause and has initially been slow – but no more; the latter has shown how a health issue in one country can spread quickly to become a health issue in every country. In just one year, climate change has triggered deep concerns and accelerated awareness as it caused sudden loss of life by fire and flood in many parts of the globe, from Canada to Turkey in terms of fire and Germany to China for floods. Covid-19 became an economic issue as workplaces shut down and some sectors experienced closure and unemployment. The issue then became politicised within countries and across countries around issues of cross-border travel and distribution of vaccinations, for example (National Geographic, 2021); further, Covid-19 created social tensions and had implications for mental health and well-being that are only now beginning to be recognised as having long-term impacts.

When it comes to the building stock, the earlier factors, along with digital transformation, changes to social mores and many more factors, combine to become drivers for retrofitting buildings that are no longer fully fit for their original purpose. However, critically, it is argued blanket assumptions that retrofits to achieve changes across the use can create as many problems as it solves. This is evidenced by the conversion of obsolete offices to flats with no outdoor space; not only do many provide poor-quality accommodation (Ferm et al., 2021), but also the impact of Covid-19 has increased the desire and health requirement for outdoor space – something which office conversions seldom offer, and at worst they can present health risks from rising temperatures.[5] This underscores once more the need for *appropriate* retrofit measures.

2.5 Resilience at the city level

According to the 100 Resilient Cities Framework (ARUP, 2014:5), 'City resilience describes the capacity of cities to function so that the people living and working in cities – particularly the poor and vulnerable – survive and thrive no matter what stresses or shocks they encounter'. As such, the definition includes physical, social and economic aspects. Their understanding derives from a historic review of the concept of resilience dating from the 1970s and the field of ecology (Walker & Cooper, 2011). This is transferable to cities when stresses and shocks threaten to cause disruption or collapse to social and physical systems. A limitation is that this conceptual framework does not include governance and power dynamics. Using this definition, resilient systems are said to feature the following seven qualities summarised in Table 2.2.

These qualities, whilst developed for the city scale, are also relevant to the building scale. The characteristics of resilient buildings have been considered in Section 2.2.

Four categories were established by the 100 Resilient Cities for urban resilience: health and well-being, economy and society, urban systems and services and leadership and strategy (ARUP, 2014:13–15). The relevance to building retrofit in

Table 2.2 Qualities of resilient systems

Quality	Characteristics
Reflective	where systems are able to accommodate inherent, increasing uncertainty with mechanisms which evolve based on emerging evidence, rather than adhering to fixed solutions based on the status quo. Using people's experiences and learning future decision-making frameworks evolve.
Robust	where systems are well conceived, built and managed so they can withstand impacts without significant damage and loss of function. Design considers future risks and potential failure, ensuring minimal safe, predictable failure occurs. Furthermore, avoiding over-reliance on any single asset is avoided as this makes the system vulnerable to catastrophic collapse.
Redundant	where systems have spare capacity to accommodate disruption, pressure or demand surges. Diversity is a key component as it increases capacity and ways of achieving different functions. Examples are distributed infrastructure networks and resource reserves. The aim is that redundancies are intentional and cost-effective, prioritised at the city scale and not an externality of inefficient design.
Resourceful	this quality is defined as the ability of people and organisations to find alternate ways to meet needs and achieve goals when experiencing shocks or stresses. This can include the capacity to predict future conditions, set priorities and plan by mobilising social, economic and physical resources. The quality of resourcefulness is vital to restore functionality of critical systems when severe conditions prevail.
Inclusive	where the need for broad consultation and engagement of all members of society is recognised. Intra-societal equity is a fundamental component of resilience. Adoption of inclusion results in shared ownership and joint vision to build resilience.
Flexible	this characteristic acknowledges change is inevitable. Flexibility can be achieved through willingness to adopt new ideas, knowledge and technologies. Importantly, there is recognition of the value of indigenous and traditional knowledge in resilience.
Integrated	alignment and integration that promotes consistency in decision-making across all scales. Effective knowledge exchange between systems enables them to function collectively and respond rapidly through effective communications.

(Source: ARUP, 2014:).

the health and well-being category is ensuring the building promotes physical and mental well-being, as discussed earlier. Examples would include providing spaces for people to enjoy proximity to nature, external green areas, or where this is not possible, use of green walls and roofs. Other examples are the use of materials that do not off gas or give off odours that might affect health. The minimal human vulnerability involves ensuing health and safety measures included in the building to avoid accidents, and the systems are robust to minimise the likelihood of systems failure, such as power outage.

Health and well-being at the city scale include the opportunity for employment for people, and at the building scale, they clearly involve some land use types such as commercial, industrial and retail that offer employment opportunities. As such, best practices in design of workstations and amenities should be adopted.

Where economy and society are concerned at the building retrofit level, the city scale categories are collective identity and mutual support, social stability and security, and availability of financial resources and contingency funds. Within organisations occupying and/or managing buildings, qualities such as inclusivity are recommended to improve resilience. Social stability and security can include providing public space where applicable and adopting design measures that enhance the safety and security of occupants and visitors. The availability of financial resources and contingency funds relates to the prudent financial management of the property by the owners to ensure effective repairs are undertaken in a timely way and regular maintenance and upgrading are provided.

For urban systems and services, the categories are reduced physical exposure and vulnerability, continuity of critical services and reliable communications and mobility. At the building scale, this involves the acknowledgement the building plays in the urban system. For example, retrofitting green roofs will contribute to attenuation of the urban heat island or can provide additional surfaces to reduce stormwater runoff into the drainage systems (Balsells et al., 2013; Wilkinson & Dixon, 2016). Where buildings are in earthquake zones, ensuring retrofits include best practice measures to minimise; collapse potential is an example of resilience. Looking at critical services in buildings: this includes power for heating, cooling and lighting; water services; and internet and communications infrastructure. Retrofits should consider having renewable energy or some renewable or emergency power capacity, storage of water and equipment that reduces water consumption with the use of recycled water where possible. A temporary emergency power supply would enable IT and communications infrastructure to work, as long as city scale infrastructure is operating. Effective property or facility management policies and plans are also needed to deliver all these operational outcomes (Siriwardena et al., 2013).

The last category at the city scale is leadership and strategy. There are three subcategories: effective leadership and management, empowered stakeholders and integrated development planning. At the resilient building retrofit scale, effective leadership and management involve having knowledgeable leaders and consultants who are aware of the local resilience issues at the city scale and how these relate to the building scale (Roostaie et al., 2019). In planning the retrofit, effective communication and discussion of all relevant issues are required for effective decision-making (Wilkinson et al., 2014). Being open to new ideas and innovations is important, and having an environment where these can be explored and debated is vital. The second sub-category, empowered stakeholders, involves the acknowledgement and recognition of all stakeholders from investors to occupiers. Consultation and education are also best practices to ensure that resilience issues are identified and discussed (Siriwardena et al., 2013). Empowered stakeholders are far more likely to deliver resilient building retrofits that will work during the retrofit and after. Integrated development planning involves effective communication across different stakeholder groups from regulatory bodies to contractors, design teams, to occupiers and visitors. In this way, relevant resilience issues are discussed and debated at all stages of the retrofit project and building life cycle. Adopting this approach will increase the likelihood of resilience being achieved in practice.

2.6 Understanding social value and the role of place

Social value is not a new concept; however, it is framed by the mores to which communities adhere, and, therefore, what constitutes social values change over time. Change in these values, which relate to ethics and morals, also impacts our expectations of individual and organisational behaviours. This has recently been clearly demonstrated, with an intensification of a search for equality, health and well-being, rights and planetary protection.

Social values set the ground rules for what governments, corporates and individuals do and expect of others. Today, widespread awareness of and concern about climate change and, more recently, Covid-19 have reshaped what is or is not acceptable socially. We expect our buildings to offer more than shelter; they need to be resilient and support health and well-being. As these social values are increasingly articulated, the need for retrofitting is required.

One example of the growth in interest in putting social values in the heart of decision-making and which, *inter alia*, affects building contract work is the UK's Social Value Act 2012. This placed a requirement on public sector procurement processes to seek not just economic benefit but also social and environmental returns. This has provided a catalyst to those seeking government contracts to develop their own corporate social responsibility policies; without them, they can no longer bid for work. One year on, an early evaluation pointed to examples of contracts, including building repair, being granted based on full triple bottom[6] criteria, but a full review found it to be patchy in uptake (Young, 2015).

As in other fields of endeavour, defining the social value within the built environment context has been difficult. Value has been, and indeed largely still is, defined in terms of economics – notably value in exchange in the marketplace.[7] To move away from this in terms of design, construction and ongoing management and to place people and the community at the heart of decisions mean re-thinking the long-held supremacy of economic return on investment. To shift the mindset to a social return on investment is not easy, with Watson et al. (2016) concluding that inconsistency of the financial metrics and data adopted leads to concerns as to the effectiveness of such measures in truly reflecting the value to the building user.

If such metrics cannot do this at the level of the building user, how can they assess social value to the wider community? UKGBC (2021), in attempting a high-level definition, attempt to do this by placing emphasis on the impact of the building on all stakeholders and by placing the needs of the local community at the heart of the design process.[8] Further, arising partly from the early experiences of Covid-19 and the resultant challenges and anticipated demand changes in the way and locus of work has come a renewed interest in placemaking. IPUT/ARUP (2020) have sought to address this through the idea of workplace making with the aim of 'recalibrating' the city environment by a deeper consideration of the spaces between buildings – such as streets, squares and green space – as integral to development.

The IPUT/ARUP report focuses on new build but translates equally to retrofit projects. If building retrofit schemes can clearly articulate or create 'permeable' buildings and places (see, e.g., Pafka & Dovey, 2017) according to the principles laid down by Jacobs (1961), then a pathway to social – as well as economic – value is created. Relating this beyond the individual building to consideration of local areas and city level is argued by Eames et al. (2014) to be critical to achieving 2050 climate targets.

In summary, it is contended that, for the fulfilment of retrofits which provide social value returns, schemes need to take an all-embracing view of who the stakeholders of a builder are and, through retrofit, can be. Adoption of, for example, or exploring multi-uses which allow and encourage permeability within and beyond buildings will allow a sense of community that can engender vibrancy and inclusivity and achieve not just a social but an economic dividend. As discussed in the section later, the retention of and investment in heritage buildings can often be core to the enhancement of social value.

2.7 Demolition versus preservation versus conservation

The debate about demolition against refurbishment and retrofit has been introduced earlier with the argument that demolition results not just in waste of resources, especially embodied carbon, but that it can have a negative impact on a sense of place. Nonetheless, even whilst supporting the notion of circularity and maximum reuse, there are times when demolition may be the only solution; the building may be structurally unsound or its location such that no new alternate purpose that is economically or socially viable is appropriate. A balanced approach but one favouring retrofit is advocated as argued by Sayce et al. (2004) and more recently by the RIBA.[9]

2.7.1 The principle

As climate change hastens, unless appropriate government-led protection schemes have taken place, risks of flood, fire or storm may render the location no longer appropriate to support human activity. Even if protection has taken place against some hazards, the prospects of extreme heat may not be capable of mitigation in some locales and may lose the building – or even the settlement – inevitable and appropriate.

In reaching decisions, environmental and social factors are critically important considerations if sustainable development goals are to be recognised and progress towards them achieved. The quantum of regulatory responses to ensure that such a 'triple bottom line' approach is adopted through planning and regulatory processes varies from country to country. Regardless of the regulatory controls and given the embodied carbon in buildings, the initial approach, from a climate change perspective, must be to seek a solution that does not involve demolition. But in the past this has normally been taken on 'single-line' economic criteria: simply which is the most profitable route to follow?

In addition to the potential, and actual, impact of the climate change factor and other environmental considerations, Covid-19 has underscored the need for buildings to provide healthy environments that promote well-being. In the commercial setting, this includes technology solutions such as ensuring high air quality; domestically, the pandemic has produced a heightened awareness of the role of outdoor space and study spaces for mental well-being and work productivity.

A further dimension in the argument is the extent to which decision-making as to the future of any asset should be determined by historic connection: should it be preserved, conserved or allowed to change to maximise economic and environmental benefit? In some cases, the case for preservation is paramount: who would look to reuse or repurpose some of the major monuments in the world? But the argument between preservation, that is retaining in existing form for posterity, and conserving, which allows for controlled and limited change, is important. Both solutions can bring economic benefit – through tourism, for example – but they can come into conflict where the move to achieve reduced carbon use is concerned. Accommodation is required in order not to destroy the cultural and historic connections whilst achieving resource use reduction. An example of how this is being attempted in the UK is in relation to minimum energy standards which are imposed on all let buildings. In the case of protected historic buildings and those in conservation areas, the standard is imposed – but only to the extent that the energy upgrade would not result in a loss of the historic features that have given rise to protection.

2.7.2 Conservation of buildings under pressure: examples

We finish this section with a few examples of where buildings have been 'saved' to a smaller or lesser extent: the first two are old examples to illustrate that the principles have long been recognised; the latter two are recent. All are from the UK, which is believed to have the oldest built stock in the world and, hence, offers many examples of where retrofit has trumped demolition.

2.7.2.1 Docklands, London, UK: industrial to residential

London Docklands was, until the early 1980s, a collection of many buildings associated with the import and export of goods and materials to and from the UK.[10] But from prosperity and growth in the 19th century, increases in the size of commercial ships meant that they could no longer gain access so far up river, and a new seaport further downstream at Tilbury led to the loss of their use by the 1970s. But whilst for years the old docks stood derelict, growth of economic, especially banking, activity overspilled the capacity of the City of London, and many parts of the old docklands became the potential for, and ultimate development of, a new business district – Canary Wharf leading to large-scale destruction of buildings and communities, despite the intention to retain where possible. But it was in the areas outside the new hub where buildings stood empty: economic values did not justify demolition and redevelopment. But, over time, many of those derelict buildings, which for lack of profitability avoided the wrecking ball, subsequently found new life, not as warehousing – but as desirable flats close to the new business centre and well-served by public investment in public transport links. A case study of some such buildings is presented in Sayce et al. (2004). In respect of sustainability, the embodied carbon in the existing structure and fabric of the buildings is very positive. Given the new uses, energy and water demands will have changed, and new services should ensure more optimum consumption.

2.7.2.2 Rodboro buildings, Guildford, UK: industrial to leisure

A further example of conservation against all odds is presented by the Rodboro Building in Guildford, UK. Reputed to be the UK's first car factory, its use changed many times before it became redundant and neglected. However, despite being listed as of historic importance, a lack of economic purpose placed the building at severe risk. It was only through determination and a sense of local history that demolition was averted, and it now trades successfully as a pub.[11] This is a further example of where full refurbishment and change of use have retained history and saved the embodied carbon and other resources; however, it was achieved for cultural and social reasons rather than any notion of the need for circulatory. Today, the desire to retain to prevent linear consumption should be an additional driver towards preferring retrofit and repurpose over demolition.

2.7.2.3 Springfield, Wolverhampton, UK: from Brewery to University 'super campus'

The old Springfield Brewery,[12] constructed in the latter part of the 19th century, is yet another example of a building that had lost value in its original use; it finally ceased operating in 1991. Subsequently, it was damaged by fire and looked set to be demolished for a housing scheme. But in much the same way as many other buildings survived, a poor economic climate rendered this unprofitable, and, with the growth in the university sector and a far greater awareness of the role of heritage, it has undergone a major programme of renewal and retrofitting, leading to its opening in 2015 as a university campus where an adjacent new building is planned as home to the National Brownfield Institute,[13] which will secure the long-term of the old brewery and create a new – but very different – community of place.

2.7.2.4 Entopia Buildings, Cambridge, UK: from Telephone Exchange to Sustainability Centre

The most recent case study offered, which demonstrates how far thinking has come, is the Entopia Building Cambridge. This is being described as a 'world-first sustainable office retrofit'[14] to home Cambridge University's Institute for Sustainability Leadership. The building was originally constructed in the 1930s as a post office for which use it became redundant, and, before the works commenced in early 2021, it would have attracted a low energy rating – such that within a few years, it would fall below minimum letting standards. The project is not just another example of reuse: it is being retrofitted to full zero carbon in use standards, appropriate for its intended new user. It will have triple glazing – still unusual in the UK, internal wall insulation and a solar power photovoltaic array on the roof. The heating will be by air source heat pump. It is designed with an intent to lower heating requirements by 75% and provide air tightness well above required levels. This combination of technologies is enabling the University to maintain, almost unchanged, the external appearance of the building, thus showcasing its heritage whilst providing an innovative retrofit intended to be a pathfinder for others.[15]

However, it should be noted that this retrofit is for an owner-occupier seeking to be a leader in sustainability and has been made possible not only through internal university funding but also by a donation of almost half the capital cost from a leading Greentech company with a further almost 25% grant from the European Development Fund.[16] The level of external grant and gift support required to achieve the aimed-for exemplary standard highlights that vision is not enough and that the business case may still not 'stack up' if a single bottom line approach is adopted.

2.8 Conclusion

This chapter has defined what is meant by a resilient building as distinct from an 'environmentally' friendly' or 'sustainable' one. Not only is retrofit a necessity due to the climate imperative, but it is also a desirable social outcome, conserving, as it does, the maintenance of place, memory and culture. It has been debated what happens when place and community are lost, for example, as in the London Docklands. The chapter has explored the extent to which retrofitting to preserve – or conserve – the social value of the building and its context is a philosophical, economic or legislative matter.

In essence, retrofits range from minor to major (Wilkinson & Remoy, 2018). Some owners may prefer to engage in an approach of little and often rather than deep or major retrofits. Owners will consider the costs, benefits and payback periods involved in the available or proposed measures. Whilst attempts have been made to assess what may constitute an optimal time frame (see, e.g., Senel Solmaz et al., 2018), it is far from a case of 'one size fits all'. But not all retrofits are designed with sustainability or resilience – a resilience retrofit must consider the full 'triple bottom line' approach. Even such retrofits vary in terms of the level of 'green' to which they aspire – with the Entopia case study presenting perhaps the deepest green approach. And this will be the requirement moving forward if we are to achieve carbon targets and start to slow climate change. The normal business-as-usual is no longer appropriate: we need acceptance that radical changes to retrofit standards are required.

It is accepted that this is not easy. It is challenging for owner-occupiers, who reap the immediate and lasting benefits, but if a building is tenanted, it has long been regarded as more difficult. However, tenant demand is changing, and if the owner wants to retain the tenants, they need to find ways to engage in measures that cause minimal disruption to occupation but that improve the tenant experience either through lower operating costs, greater comfort and/or improved reputation through the occupation of an improved building. This is often the case with residential social landlords, for whom tenant well-being is a high priority, although as far as possible, they may well choose to undertake energy improvements 'in-cycle', meaning other renewals (e.g. kitchen/bathroom fittings require replacement). In respect of resilience, retrofits may enable owners to enhance building resilience to shocks and stresses as our knowledge and understanding of them evolves.

Retrofit will always be required: whereas a building envelope can normally be expected to outlast its economic usefulness in its existing use, building services typically have a life cycle of only around 25 years or so before replacement is required – and they may become inefficient in terms of performance long before this. At this stage, there may be a need to vacate part or all of a building, and it can be a good time to consider other improvements and/or repairs to the building fabric and envelope. Other more significant sustainability and resilience options can, and should, be considered at this 'in-cycle' stage.

Some retrofit measures will also enhance the capital value of the building; however, the gain may be less than the cost, and this continues to be a disincentive for many.

But protecting the value and preventing the installation of retrofit measures, although they may enhance capital value, will not necessarily produce a capital gain greater than the cost. If this is the case, they may face a very real risk of the building becoming stranded in value terms (for a discussion of this, see Muldoon-Smith & Greenhalgh, 2019).

Finally, the chapter argues that retrofitting to preserve – or conserve – the social value of the building and its context is a philosophical, economic or legislative matter. Whilst this has long been the case, and illustrations of building survival have been quoted, heightened awareness of climate change and the lens of Covid-19 places heightened responsibility on all building owners to seek to create truly remarkable retrofits which meet the very highest environmental ambitions.

Notes

1 For example, it is claimed that many new homes built in England since 2018 already need cooling systems installed to combat rising summer temperatures. www.telegraph.co.uk/environment/2021/07/10/homeowners-face-9000-bills-stop-new-builds-overheating/
2 www.oxfordlearnersdictionaries.com/definition/american_english/renovate
3 For information on the Resilient Cities Agenda, see www.arup.com/perspectives/publications/research/section/city-resilience-index
4 realassetinsight.com/2021/07/12/riba-demolitions-should-be-stopped-to-lower-emissions/#:~:text=The%20Royal%20Institute%20of%20British,net-zero%20targets%20by%202050.&text=Every%20year%2050%2C000%20buildings%20are,of%20the%20UK's%20total%20waste.

5 See www.theguardian.com/society/2021/aug/01/converted-offices-pose-deadly-risk-in-heatwaves-experts-warn
6 The Triple Bottom Line, which balances economic, social and environmental factors in decision-making is widely attributed to Elkington (1997).
7 IVSC (2016) provides the most widely adopted definition of market value as being Market Value is the estimated amount for which an asset or liability should exchange on the valuation date between a willing buyer and a willing seller in an arm's length transaction, after proper marketing and where the parties had each acted knowledgeably, prudently and without compulsion.
8 See www.ukgbc.org/ukgbc-work/framework-for-defining-social-value/
9 See comment on Section 2.3.
10 For a brief history, see www.royaldocks.london/articles/a-history-of-the-royal-docks
11 This case study is written up in Sayce et al., 2004 and a short history can also be found at www.guildfordsociety.org.uk/rodboro.html.
12 For a brief history, see www.wlv.ac.uk/university-life/our-campus/springfield-campus/heritage-of-springfield/
13 https://investwm.co.uk/2020/12/14/17-5m-national-brownfield-institute/
14 www.cisl.cam.ac.uk/about/entopia-building
15 www.theconstructionindex.co.uk/news/view/cambridge-refurb-project-to-be-the-deepest-shade-of-green
16 www.theconstructionindex.co.uk/news/view/cambridge-refurb-project-to-be-the-deepest-shade-of-green

References

Ancient History, 2020. Heraclitus of esphesus. Available from https://www.ancient.eu/Heraclitus_of_Ephesos/. Accessed 6 January 2021 and 15 June 2021.
ARUP, 2014. *City Resilience Framework*. ARUP group Ltd., London. Available from https://www.arup.com/projects/city-resilience-index. Accessed 16 May 2022.
Ashworth, A., 1997. *Obsolescence in Buildings: Data for Life Cycle Costing*. The Chartered Institute of Building, London, Report No. 74.
Ballinger, N., 2020. *Building Tune-Up Accelerator Program Final Technical Report* (No. DOE-SEATTLE-07556). City of Seattle. https://doi.org/10.2172/1630737.
Balsells, M., Barroca, B., Amdal, J. R., Diab, Y., Becue, V. and Serre, D., 2013. Analysing urban resilience through alternative stormwater management options: Application of the conceptual spatial decision support system model at the neighbourhood scale. *Water Science and Technology*, 68(11), pp. 2448–2457.
Baum, A., 1993. Quality, depreciation, and property performance. *Journal of Real Estate Research*, 8(4), pp. 541–565.
Bowie, N., 1982. Depreciation: Who hoodwinked whom?" *Estates Gazette*, 262, pp. 405–411.
Brand, S., 1997. *How Buildings Learn: What Happens After They're Built*, Revised edition. Phoenix Illustrated, London, UK.
Bryson, J. R., 1997. Obsolescence and the process of creative reconstruction. *Urban Studies*, 34(9), pp. 1439–1458.
Chanter, B. and Swallow, P., 1996. Maintenance organisation. In *Building Maintenance Management*. Blackwell Science, London.
Chen, C. J., Juan, Y. K. and Hsu, Y. H., 2017. Developing a systematic approach to evaluate and predict building service life. *Journal of Civil Engineering and Management*, 23(7), pp. 890–901.
City of Melbourne. 2021. What is a building retrofit? Available from https://www.melbourne.vic.gov.au/business/sustainable-business/1200-buildings/building-retrofit/Pages/building-retrofit.aspx. Accessed 16 May 2022.

Cook, S. J. and Golton, B., 6–9 November 1994. *Sustainable Development Concepts and Practice in the Built Environment – A UK Perspective*. CIB TG 16. Sustainable Construction, Tampa, FL.

Cowan, P., 1970. *Obsolescence in the Built Environment: Interim Reports*. Joint Unit for Planning Research, London.

Crawford, R., 2011. *Life Cycle Analysis in the Built Environment*. Routledge, New York.

de Wilde, P., 2019. Ten questions concerning building performance analysis. *Building and Environment*, 153, pp. 110–117.

Eames, M., Dixon, T., Lannon, S. C., Hunt, M., De Laurentis, C., Marvin, S., Hodson, M., Guthrie, P. and Georgiadou, M. C., 2014. *Retrofit 2050: Critical Challenges for Urban Transitions*. Cardiff University, Cardiff.

Ellison, L. and Sayce, S., 2007. Assessing sustainability in the existing commercial property stock: Establishing sustainability criteria relevant for the commercial property investment sector. *Property Management*, 25(3), pp. 287–304.

Ferm, J., Clifford, B., Canelas, P. and Livingstone, N., 2021. Emerging problematics of deregulating the urban: The case of permitted development in England. *Urban Studies*, 58(10), pp. 2040–2058.

Golton, B. L., 1989. Perspectives of building obsolescence. In *Land and Property Development: New Directions*. E. & F. N. Spon, London.

Grover, R. and Grover, C., 2015. Obsolescence – a cause for concern? *Journal of Property Investment & Finance*, 33(3), pp. 299–314.

Gustafsson, M., Dipasquale, C., Poppi, S., Bellini, A., Fedrizzi, R., Bales, C., Ochs, F., Sié, M. and Holmberg, S., 2017. Economic and environmental analysis of energy renovation packages for European office buildings. *Energy and Buildings*, 148, pp. 155–165.

IPUT/ARUP, 2020. Making place: Global research report the recalibration of work, life, and place. Available from www.arup.com/perspectives/publications/research/section/making-place-the-recalibration-of-work-life-and-place.

Jacobs, J., 1961. *The Death and Life of Great American Cities*. Vintage Books, New York, NY.

Jennings, J. B., Vugrin, E. D. and Belasich, D. K., 2013. Resilience certification for commercial buildings: A study of stakeholder perspectives. *Environment Systems & Decisions*, 33, pp. 184–194,

Jump, D., Denny, M. and Abesamis, R., 2007. Tracking the benefits of retro-commissioning: M&V results from two buildings. Proceedings of the 2007 National Conference on Building Commissioning, May 2–4, Chicago. Available from www.peci.org/ncbc. Accessed 15 May 2015.

Khalid, G., 1993. Hedonic price estimation of the financial impact of obsolescence on commercial office buildings (PhD), University of Reading.

Kishk, M. O. and Al-Hajj, A., 2000. Handling linguistic assessments in life cycle costing-a fuzzy approach. The Construction and Building Conference of the RICS Research Foundation (COBRA 2000). University of Greenwich.

Kontokosta, C. E., Spiegel-Feld, D. and Papadopoulos, S., 2020. The impact of mandatory energy audits on building energy use. *Nature Energy*, 5(4), pp. 309–316.

Lichfield, N., 1968. Economics in town planning: A basis for decision making. *The Town Planning Review*, 39(1), pp. 5–20.

Mansfield, J. R. and Pinder, J. A., 2008. "Economic" and "functional" obsolescence. *Property Management*, 26(3), pp. 191–206.

Moazami, A., Carlucci, S. and Geving, S., September 2019. Robust and resilient buildings: A framework for defining the protection against climate uncertainty. In *IOP Conference Series: Materials Science and Engineering* (Vol. 609, No. 7, p. 072068). IOP Publishing, Bristol.

Muldoon-Smith, K. and Greenhalgh, P., 2019. Suspect foundations: Developing an understanding of climate-related stranded assets in the global real estate sector. *Energy Research & Social Science, 54,* pp. 60–67.

Nanayakkara. K., Wilkinson, S., & Ghosh, S., 2021. Future office layouts for large organisations: workplace specialist and design firm's perspective. *Journal of Corporate Real Estate.* doi: https://doi.org/10.1108/JCRE-02-2020-0012.

National Geographic, 21 April 2021. Travelers are crossing borders for vaccines. Is that okay? Available from www.nationalgeographic.com/travel/article/traveling-across-borders-for-a-vaccine.

Pafka, E. and Dovey, K., 2017. Permeability and interface catchment: Measuring and mapping walkable access. *Journal of Urbanism: International Research on Placemaking and Urban Sustainability, 10*(2), pp. 150–162.

Phillips, R., Troup, L., Fannon, D. and Eckelman, M. J., 2017. Do resilient and sustainable design strategies conflict in commercial buildings? A critical analysis of existing resilient building frameworks and their sustainability implications. *Energy and Buildings, 146,* pp. 295–311.

Pinder, J. and Wilkinson, S. J., 2001. Measuring the obsolescence of office property through user-based appraisal of building quality. CIB World Building Congress: Performance in Product and Practice, 2nd–6th April, Wellington, New Zealand.

Pourebrahimi, M., Eghbali, S. R. and Roders, A. P., 2020. Identifying building obsolescence: Towards increasing buildings' service life. *International Journal of Building Pathology and Adaptation.* doi:10.1108/IJBPA-08-2019-0068.

Reed, R. and Warren-Myers, G., January 2010. Is sustainability the 4th form of obsolescence? In *PRRES 2010: Proceedings of the Pacific Rim Real Estate Society 16th Annual Conference* (pp. 1–16). Pacific Rim Real Estate Society (PPRES), Wellington, New Zealand.

RICS (Royal Institution of Chartered Surveyors), 2018. Depreciated replacement cost method of valuation for financial reporting guidance note. Available from www.rics.org/uk/upholding-professional-standards/sector-standards/valuation/depreciated-replacement-cost-method-of-valuation-for-financial-reporting/.

The Rockefeller Foundation, 2022. 100 resilient cities. Available from https://www.rockefellerfoundation.org/100-resilient-cities/. Accessed 16 May 2022.

Rocky Mountain Institute, 2020. The multiple benefits of deep retrofits: A toolkit for cities. Available from www.c40knowledgehub.org/s/article/The-Multiple-Benefits-of-Deep-Retrofits-A-toolkit-for- cities?language=en_US.

Roostaie, S., Nawari, N. and Kibert, C. J., 2019. Sustainability and resilience: A review of definitions, relationships, and their integration into a combined building assessment framework. *Building and Environment, 154,* pp. 132–144.

Salway, F., 1986. *Depreciation of Commercial Property: CALUS Research Report: Summary.* College of Estate Management. Reading, MA.

Sayce, S., Walker, A. and McIntosh, A., 2004. *Building Sustainability in the Balance: Promoting Stakeholder Dialogue.* Estates Gazette, London.

Senel Solmaz, A., Halicioglu, F. H. and Gunhan, S., 2018. An approach for making optimal decisions in building energy efficiency retrofit projects. *Indoor and Built Environment, 27*(3), pp. 348–368.

Siriwardena, M., Malalgoda, C., Thayaparan, M., Amaratunga, D. and Keraminiyage, K., 2013. Disaster resilient built environment: Role of lifelong learning and the implications for higher education. *International Journal of Strategic Property Management, 17*(2), pp. 174–187.

Smith, M., Whitelegg, J. and Willmias, N., 1998. *Materials Intensity in the Built Environment.* Greening the Built Environment Earthscan Publications Ltd, London.

Thomsen, A. and Van der Flier, K., 2011. Understanding obsolescence: A conceptual model for buildings. *Building Research & Information*, 39(4), pp. 352–362.

Van Dronkelaar, C., Dowson, M., Burman, E., Spataru, C. and Mumovic, D., 2016. A review of the energy performance gap and its underlying causes in non-domestic buildings. *Frontiers in Mechanical Engineering*, 1, p. 17.

Walker, J. and Cooper, M., 2011. Genealogies of resilience: From systems ecology to the political economy of crisis adaptation. *Security Dialogue*, 42(2), pp. 143–160.

Watson, K. J., Evans, J., Karvonen, A. and Whitley, T., 2016. Capturing the social value of buildings: The promise of Social Return on Investment (SROI). *Building and Environment*, 103, pp. 289–301.

Wikipedia Rochford Hall, 2021. Available from https://en.wikipedia.org/wiki/Rochford_Hall. Accessed 15 June 2021.

Wilkinson, S. J., 2012. Conceptual understanding of sustainability in Australian property firms. European Real Estate Conference, 13–15 June 2012, Edinburgh, Scotland.

Wilkinson S. J. and Dixon, T., 2016. *Green Roof Retrofit: Building Urban Resilience*. Wiley-Blackwell, Chichester and West Sussex, UK

Wilkinson, S. J. and Remoy, H. (Eds.), 2018. *Building Resilience in Urban Settlements through Sustainable Change of Use*. Wiley-Blackwell, Chichester.

Wilkinson, S. J., Remoy, H. T. and Langston, C., 2014. *Sustainable Building Adaptation: Innovations in Decision-making*. Wiley-Blackwell, Oxford UK.

Young, D., 2015. *Social Value Act Review*. Cabinet office, London.

3 An inadequate building stock

Sarah Sayce and Sara Wilkinson

3.1 Introduction

This chapter describes the challenges for the existing building stock set by climate change and the recent impacts of Covid-19. The many challenges are varied and are explored in more detail in later sections. However, the inadequacies most explored here are connected with the building age and design limitations, leading to associated energy inefficiency, resource consumption and greater exposure to other climate risks. The challenge of the stock of non-resilient and inappropriate buildings is currently not being met. Retrofitting for low energy and low carbon, which have been seen to be the priority measures for most governments, is only part of the picture: water conservation, flood protection and the ability to withstand fire are increasingly important criteria for resilient, sustainable buildings. Recent disasters, such as the Grenfell Tower fire in London, which involved retrofitting a residential building with external thermal cladding shortly before a deadly fire broke out in 2017, stand testament to the inadequacy of some contemporary design materials and weak compliance with standards required now, let alone moving forward.

Further, as the health and well-being agenda gains traction and research connects chronic and fatal conditions, such as asthma, with pollution and building defects, the quality of buildings is further brought into focus. Again, this is not a new issue: damp and overcrowded buildings prevalent in the slums of the 19th century in recently industrialised countries have long been recognised as harbingers of sickness and exacerbators of existing pre-conditions. More recently, the advent of air-conditioning systems and central heating systems led to increases in Legionnaires' disease, which flourishes in warm, stagnant water; whilst the pressure to increase the density of buildings and cut down ventilation to maintain warmth in cold climate led to the recognition of sick building syndrome.

In many ways, the length of time that it has taken to raise the issue of connecting human health and well-being with building design and use is surprising; it is no wonder that air quality and healthy buildings are included in many sustainability ratings and have formed the rationale for WELL certification[1] which was first launched only in 2016. Then, in 2020, a fatal and hitherto unacknowledged risk came to the fore; the transmission of disease within buildings through heating

DOI: 10.1201/9781003023975-4

ventilation and air conditioning (HVAC) systems. The 2020 Covid-19 global pandemic showed transmission of the virus through HVAC systems was possible. This risk in some jurisdictions led to office workers being mandated to work from home to reduce the risk of exposure and transmission of the virus; in other countries, this was not required but encouraged.

Working from home has worked well for some staff and employers. There are ongoing discussions and debates about the viability of the return to work in offices post-pandemic and discussions of the benefits to stakeholders of allowing some work from home. Working from home can be logistically challenging for some people or may lead to deteriorating mental health (Evanoff et al., 2020; Palumbo, 2020). Even if workers return to offices, the type of space and how it operates will likely change, accelerating trends already identified (Harris, 2020). In the short term, at the very least, this may lead to large vacancy levels in the office sector as leases terminate and leaseholders seek smaller demised spaces to operate from, and in the longer term, buildings which cannot 'flex' to meet changing demand may become economically 'stranded'. Almost inevitably, there will be growth in 'inadequate stock'.

Similarly, the pandemic has caused humans to evaluate what they require from their homes: green space, good ventilation and study space separate from general living areas. This is causing a 're-think' among homes designers and a sharp change in demand – both in specification and in location. The demand for gardens and home offices in suburban areas has risen.[2] As a result, dwellings, including many conversions from redundant offices, are now often no longer seen as fit for purpose. In turn, this could affect city development (Avasarala et al., 2020) as planners and policymakers try to reconcile changing, currently not fully understood trends.

Last, whilst the chapter does examine how buildings can, and do become inadequate, it emphasises that inadequacy extends beyond just the process of ageing. There is often a mismatch of buildings to the social needs of communities: be that through the shift in transport means, such as the introduction of electric (or hydrogen) and, in time, possibly autonomous vehicles, and the need to accommodate greater densities of population and to feed them locally. By outlining the drastic, radical changes needed, the chapter concludes that collectively this creates a multifaceted challenge to adapt and improve existing buildings.

3.2 Profiling the existing building stock

3.2.1 Age

Dean et al. (2017) predicted the total stock of buildings in developed countries increases by an estimated net 1%–2% each year. This statement was published in The Global Status Report 2016 prepared by the Global Alliance for Buildings and Construction for the 22nd Conference of Parties (COP22) to the United Nations Framework Convention on Climate Change (UNFCCC). It is also estimated that the total stock will have doubled by 2050 from 2020 levels (Dean et al, 2017.

Partly this is in response to growing populations, but not completely as other social and legislative factors come into play which determines both the demand for and usability of the building.

Even allowing for demolitions, a large majority of buildings are aged – and most will still be standing in 2050 – when many nations are aiming to be zero carbon. In addition, many buildings are 'hard to treat' to improve to modern specifications and performance at an economically viable cost. Consequently, they may be ill-suited for adaptation to meet occupational requirements in a digital, zero-carbon world. The rate of demolition and redevelopment will vary with economic cycles, social trends and cost/economic return ratios. Still, the lower the economic returns of redevelopment, the longer stock, counter-intuitively, may stay standing (Sayce et al., 2004). This is because obsolete buildings, which have outlived their current useful purpose, may be sited in areas with a low value for redevelopment. This can lead to dereliction or disrepair until development becomes viable. Adaption tends to be in areas of high land values and where density and height of allowed development permit conditions conducive for the rapid redevelopment of existing stock. Whilst most buildings are designed for functional, technical lifespans in mind, the pace of change of planning policy, social factors and economics are typically the defining factors, unless the building is protected due to imposition of heritage status.

This confirms the argument posited by Ball (2003:188) that redevelopment is 'most likely to be influenced by contemporary economic, regulatory and other non-age factors – i.e. where users want to be located and what regulators wish to preserve and limit'. More recently, Scott (2019) attributes the decision to redevelop for 'lucrative' gains as prevalent in what Ehrenreich (2016) has dubbed 'third-wave' capitalist cities characterised by high gains for some and severe losses for others – all of which are markers of an unequal society, financially and socially. An example of this is the City of London, where planning has progressively allowed for taller buildings, resulting in some office blocks being developed three times in some 70 to 80 years. This practice highlights a discrepancy between an economic life which is often below the original design life.

The result of the mixed menu of factors determining building life, many of them independent of structural integrity, provides a challenge to retrofit with those buildings most in need of investment to bring them to the standards required for resilience into the future often lay within the ownership of those least able to afford to make that investment. This theme is picked up again later.

3.2.2 *Energy efficiency and carbon emissions*

It has been widely acknowledged that buildings stocks are a very significant source of carbon: the level varies from country to country and over time. However, the figure is incompatible with almost all countries' low or zero economies. Figure 3.1 shows a wide disparity when assessed in terms of Energy Performance Certificates (EPCs). These figures, however, should not be taken at face value. For example,

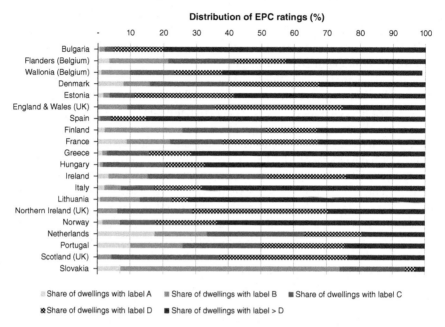

Figure 3.1 Comparative Energy ratings of Domestic Stock across selected European Countries

Source: BPIE based on data derived from various sources including https://gitlab.com/hotmaps/building-stock/-/blob/master/README.md

the methodology for inspecting and assessing EPCs vary, meaning that A grade EPC is not consistent between countries. There are also inconsistencies at the point at which certification take place. In the UK, one is required before letting or sale, so it is highly skewed to existing stock, whereas in Slovakia, which appears to be performing well, most certificates relate to new buildings.

The 2019/2020 English Housing Survey (Gov. UK, 2020) stated the energy efficiency of the English housing stock continued to improve, albeit slowly. In 2019, the average SAP rating of English dwellings was 65 points, up from 63 points in 2018. This was evident in all tenures apart from social housing, where there was no significant increase. However, overall, the UK social sector is more energy-efficient than the private sector. In the social rented sector, the majority of dwellings (61%) were in EER bands A to C, compared to 38% of private rented sector dwellings and 36% of owner-occupied dwellings. In dwellings, a C rating or better is now regarded as the 'baseline' for a 'good' level of efficiency, and it rather underscores the point that there is a significant difference between tenures. People in financially compromised positions are more likely to be concentrated in the private rented sector, further exacerbating occupants' issues around paying for heating, rent and food. Figure 3.2 shows energy efficiency ratings (EER) for existing and new domestic buildings in England from Jan to March 2021.

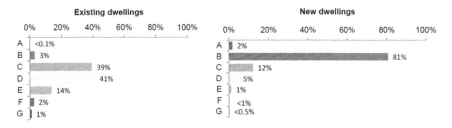

Figure 3.2 Energy efficiency ratings (EER) for existing and new domestic buildings in England from January to March 2021

(Source: Gov UK, 2020)

The energy efficiency of a building is often a function of its age. It is estimated that within Europe, which has the oldest building stock globally, fewer than 5% of buildings are 'energy efficient': far less are zero carbon. Indeed, the European Commission estimates that in most EU countries, half of the residential stock was built before 1970, before the first thermal regulations were introduced; the UK has probably the oldest stock with an estimated 31% of dwellings pre-dating 1945.[3] The date at which the introduction of minimum energy standards in building regulations and codes varies. The UK, for example, was an 'early' adopter, with compulsory thermal performance standards for new buildings introduced in Building Regulations in 1984. However, it has tended to be countries with high wealth and harsh winters that could introduce such measures without impeding developments where these were needed. In warmer climes, or those with poor regulatory frameworks, or with plentiful access to cheap power sources, the issue of thermal performance was less likely to be mandated; for example, other countries like Australia did not mandate minimum energy performance in residential buildings until 2005 (Wilkinson, 2011).

In addition to the age of buildings and the date at which minimum standards were mandated, a further contributory factor in the pursuit of energy-efficient buildings has been challenges to require energy upgrades at the same time as alteration, extension or refurbishments works have taken place and, in some cases, poor enforcement regimes. The result is that, as discussed in later chapters, upgrading may have to be incentivised through fiscal or grant initiatives or linking occupation or upgrades to mandating standards for continued investment stock, such as has happened in the UK and the Netherlands.

3.2.3 Fire standards and performance

Buildings are required to meet minimum standards of fire resistance and fire spread at the point of construction through adherence to building regulations. Standards change over time; for example, building materials and designs once deemed acceptable are found to be inadequate, and standards change. In addition, alterations to buildings plus wear and tear mean that original designs no longer exist or perform as designed. Occupational densities may also change, and sometimes this is known to building owners or managers. At other times, over-occupation, for example, is not known to owners or managers. Where buildings are inadequate in respect of fire performance, action must be taken.

3.2.4 Building management inefficiency

Buildings will only perform as designed if they are well managed. Different occupants behave very differently in respect of how they operate buildings (Wilkinson & Kalejska-Jonsson, 2019). This means that water and energy consumption levels can vary considerably between parties occupying the same space, for example, and over predicted usage. The knowledge of the building facility manager is another key factor in the operational efficiency of commercial stock (Fortino et al., 2012). Older buildings tend to have systems predating the sophisticated, so-called 'smart' technology specified in new building design to ensure optimal operational efficiency regarding water and energy (Corna et al., 2015; Zhang et al., 2013). Smart technology uses computerised systems and sensors for optimum building operation.

3.2.5 Unhealthy buildings

Much has been written about unhealthy buildings. Sick building syndrome (SBS) is a condition whereby building occupants suffer from symptoms of illness or become infected with chronic disease from the building they live or work in. The outbreaks may arise from inadequate or inappropriate cleaning (Passarelli, 2009). SBS describes occupant concerns about defects in the construction materials or the construction process and/or inadequate maintenance (Passarelli, 2009). Symptoms can worsen with the time spent in the building, and symptoms can decrease or disappear when people leave the building. Exposure to mould can be a problem (Strauss, 2009; Terr, 2009). A 1984 World Health Organization (WHO) report suggested up to 30% of new and retrofitted buildings globally may be subject to complaints about poor indoor air quality (Environmental US, 1991). Other causes have been attributed to contaminants produced by 'off gassing' of some building materials, volatile organic compounds (VOCs), moulds (Strauss, 2009; Terr, 2009) improper exhaust ventilation of ozone (a by-product of some office machinery), and light industrial chemicals used within, or lack of adequate fresh-air intake/air filtration (Passarelli, 2009). Where existing buildings have issues around occupant health, action is required to remedy the issue.

3.2.6 Excessive operating costs

Another feature of our inadequate stock is that they cost more to operate compared to new, efficient stock (Roberts, 2008; Dobson et al., 2013). Reasons can include original building predates regulations for insulation, or building services such as heating, cooling and lighting are aged and inefficient compared to contemporary equipment. Smart building technology to optimise performance is relatively recent, with the term defined in 2014 (Buckman et al., 2014). Retrofit provides an opportunity where building services are inefficient and/or dated to reduce energy and water consumption and lower operating costs.

3.2.7 Economic obsolescence

Mansfield and Pinder (2008) found depreciation in commercial buildings value arose from three forms of obsolescence – physical, economic and functional. The economic decline in utility ultimately lowers the value and may stimulate owners to retrofit or convert the property to optimise value. Whilst economic obsolescence may not be associated with the age or condition of the property (Mansfield & Pinder, 2008), physical changes may be needed to create the changes needed to increase capital and/or rental value.

3.2.8 Why not demolish?

The question of demolition arises when any of the factors discussed in this section are raised. There are several reasons why demolition is not always the best option. The reasons can be classified as environmental, economic and social. For example, the existing structure represents a huge amount of embodied energy, and a retrofit allows much of that embodied energy to be retained (Azari & Abbasabadi, 2018). Socially, over time buildings become associated with a sense of place; memories of events, and human connection (Kirby & Kent, 2010; Bullen & Love, 2011). Finally, the economic drivers for retrofit can be that the project is completed sooner, at lower construction cost and income revenue can be achieved sooner (Medal & Kim, 2020).

3.3 Adaptations required to meet the climate crisis

3.3.1 Water conservation

Fresh potable water is an issue in many cities globally, particularly in the Middle East, South Africa, and parts of the US (notably California) and Australia. These are regions with limited rainfall and growing populations. Historically, water conservation adopted a very low-tech approach, with properties each having dams to capture rainwater. However, current development over the last 25 years has increasingly concentrated populations into higher rise, higher density settlements. Our lifestyles have embraced machines to wash clothing and cooking utensils; typically, these machines use more water, potable water, to complete these functions. Some eco-products are available but typically have a price premium.

Indeed, so bad have water shortages become because of climate change and more affluent lifestyles in some areas; estimates state some 4 billion people are living in water scarcity on a daily basis (Mekonnen & Hoekstra, 2016). In South Africa, for example, after three years of drought, in 2018, Cape Town was expected to run out of water. Whilst enforced behavioural change did avert the worst impacts, it is predicted that due to climate change, there is a strong possibility that such a scenario will be repeated (Pascale et al., 2020). Cape Town is far from being a unique example.

There are ways to reduce operational water consumption. Adopting the mantra; reduce, reuse and recycle, it is possible to reduce consumptions by specifying spray taps in bathrooms and kitchens. These taps use less water for cleaning functions. Encouraging people to shower rather than take baths is another way to reduce water consumption. Planting gardens with native plants instead of imported species require less watering to maintain the plants. For residential developers, offering purchasers kitchen and washing equipment with low water consumption is another way to reduce use. Finally, ensuring homebuyers fully understand how to operate their homes to optimise reduced consumption is essential (Wilkinson & Zajelska-Jonsson, 2021; Wilkinson & Kalejska-Jonsson, 2019; Wilkinson et al., 2013), as if not, the potential gains will not be realised.

Another option developers have is to retrofit technologies to reuse and recycle water. For example, greywater, which is water discharged from kitchen and bathroom sinks, showers and baths, and washing machines, can be used to water gardens, plants and lawns, green roofs and walls. It requires designers to reroute waste pipes to discharge to garden areas. This can be done at an individual house scale or medium and high-density apartment or precinct block scale. It can be adopted in commercial, retail and industrial building stock. The amounts of water reused and recycled will vary depending on the type and density of development.

There is the potential to incorporate rainwater harvesting tanks in retrofit projects. Rather than rainwater discharging into the sewer system, on-site tanks capture the rainwater and store it for later use. These tanks are often located below ground for larger projects or in above ground tanks for smaller scale density developments. Careful planning in the retrofit design stage can accommodate these technologies in many cases.

3.3.2 Flood protection

In low lying areas, property is becoming increasingly vulnerable to sea-level rise. According to the US climate centre NOAA (Lindsey, 2020), sea level has risen 20 cm already this century in 20 years, with a further 80 cm predicted to 2100. Coupled with king tides and surges, this means more existing buildings will be vulnerable to experiencing flooding. With the increased incidence of inundation, some locations will find it impossible to get flood protection cover on home insurance or building policies. Insurance councils globally are assessing climate risk and insurance premiums, and adjustments will reflect the likelihood and severity of predicted events (Bhattacharya Mis et al., 2018. This will ultimately impact on value. In extreme cases, places will be abandoned as sea-level rises may make ongoing inhabitation untenable.

There are numerous ways flood protection may be achieved. At the regional scale, governments can design flood protection through sea walls and other major infrastructure measures to protect land and property. Though effective, these measures are long term, expensive and beyond the scope of small and even large property developers in retrofit projects. However, at the building scale, there are measures that can be taken (Wilkinson et al., 2014). In retrofit, avoiding basements and below ground parking, elevating the ground floor so that flood water can pass under and through the property are examples of flood protection measures that may not be obvious to most people (Bhattacharya Mis et al., 2018, Lamond et al., 2019).

Other approaches acknowledge that flooding will occur and when it does, fast recovery is the goal. Here adoption of specifications and materials that dry out rapidly without spoiling may be the preference. For example, using industrial style concrete wall finishes ensures less damage is suffered and that drying out is quicker. This makes a recovery, post-inundation, more economical than using traditional plasterboard and porous wallpaper finishes which are spoiled in floods.

3.3.3 Fire resilience

The need for fire resilience comes from various sources. For example, in some countries and locations, there are dangers from wildfires or bushfires. This is where forest and natural environments spontaneously combust, and fire is spread by wind. Depending on the direction and speed of the wind, large settlements can be susceptible to fire risk. The changing climate is increasing the frequency and intensity of these fire events. In late 2019 and early 2020, Australia experienced catastrophic bushfires in New South Wales and Victoria over several weeks. Older stock will not meet contemporary minimum fire standards, and retrofit will involve upgrades. Local building codes set out minimum standards in respect of fire protection and best practices in terms of siting trees and other combustible materials near buildings. Having stores of water on site will help with fighting fires also and is complementary to the reuse and recycling of water and storage of rainwater noted in Section 3.2.1. Clearly, with existing property, the current fire protection and materials used should be considered along with the current and future standards likely to be introduced. Future proofing retrofitted buildings by going beyond minimum contemporary standards is highly advisable whenever possible.

One of the tragic issues faced in the June 2017 Grenfell Tower fire in Kensington, London, was that retrofitted external cladding panels installed to increase thermal performance were flammable (Grenfell Tower Inquiry, 2019). Tragically, 72 residents of the 23-storey high rise apartment building died in the fire. The original building needed upgrading thermally to meet the 2017 energy efficiency minimum standards on the UK Building Regulations, and the cladding was deemed able to meet that requirement. Enquiries concluded the design was flawed, and combustible materials in the external cladding enabled a kitchen fire to spread to the entire building. Combined with the rapid fire spread and toxic fumes, residents were unable to evacuate the building safely. The property management and fire brigade were criticised in the report on the tragedy (Grenfell Tower Inquiry, 2019). This highlights the critical need to balance environmental goals of reduced operating energy and lower greenhouse emissions with health and safety and, importantly; effective property management. Due diligence, investigating all potential risks must be undertaken. Furthermore sometimes, manufacturers' product information might be inadequate.

Another method of fire protection is called active fire protection. Fire alarms and sprinkler systems are examples of active systems, whereas external wall cladding is an example of passive fire protection. Older buildings tend to predate the inclusion of active firefighting technologies in the original specification, and there could be a reliance on passive systems despite a building being used in different ways than originally designed or new upgrades/maintenance. Both measures can be retrofitted into buildings depending on clients' wishes, building codes and standards and the clients' budget. Technically it is very rare that a building cannot be retrofitted for fire resilience. Critically, qualified fire engineering advice is required to ensure the retrofit is building code/regulation compliant and the occupants' safety is paramount.

3.3.4 Health and well-being – pollution

There are several factors to consider regarding health and well-being in building retrofit. In the existing building it is possible, materials previously specified are now deemed deleterious to human health. Asbestos cladding is a good example. Originally specified as a fire-retardant cladding material in the 19th century, asbestos was widely used in industrial and residential buildings globally. Overtime, medical research linked some forms of asbestos to cancer in humans. The minute fibres in the asbestos become more friable over time. As the material ages and possibly sustains impact damage, the fibres are released into the air and potentially inhaled by building occupants. Factors such as frequency of exposure and other compounding factors, including lifestyle choices, result in varying rates of asbestosis, an often-fatal cancer. Surveys of existing buildings are essential to detect any deleterious materials, such as asbestos, and then if found, arrangements for safe disposal should be instigated (World Health Organization, 2010. It should be noted that there are different types of asbestos; and it can be removed or, where it is in an acceptable condition, managed for example see the UK regulations Control of Asbestos Regulations 2012 (Books, 2013).

Other materials in existing buildings may contain VOCs, which are to be avoided in specifications for retrofit. VOCs 'off gas' over time following installation. VOCs are a large group of chemicals that are found in many products used in buildings and maintenance. Off gassing involves the release of odours from the VOC materials, which are inhaled, causing deleterious health effects. Common VOCs include acetone, benzene, ethylene glycol, formaldehyde, methylene chloride, perchloroethylene, toluene and xylene (Keywood et al., 2016). Clearly, an audit of existing materials is essential, followed by safe removal where found. The new retrofit specification should ensure no VOCs are introduced.

Wherever possible, retrofit designs should enable good natural cross ventilation of property to ensure good rates of fresh air are provided for occupants. Proximity to sources of pollution such as motorways and airports should also be factored into retrofit projects to ensure the highest standards in respect of pollution mitigation and abatement are adopted.

3.3.5 Deleterious materials and building defects

As mentioned earlier, previously materials now considered deleterious were legal and used in buildings. As well as asbestos, another example is the use of lead in piping. Depending on the quality of water, lead can be dissolved and then ingested by humans. Over time extended exposure can contribute to lead poisoning and lead-based cancers. Depending on the age and location of the property, different preferences exist in respect of specifications, and the professional team should ensure a thorough audit and visual inspection of all aspects of the existing building is undertaken to identify any deleterious materials.

Another issue to consider is the existence of building defects which could affect the technological and economic viability of the proposed retrofit. For this, a Technical Due Diligence survey, undertaken by a professionally qualified chartered Building Surveyor is essential (RICS, 2020). This professional is educated and trained in the inspection of existing buildings and the diagnosis of building defects. Over time all buildings age and deteriorate, however, the rates of which are affected by the design and specification adopted, as well as the amount and quality of maintenance.

Defects to consider are evidence of any building movement, usually manifesting in cracking and/or displacement. The cause of the movement should be identified, and whether it is ongoing or historic. Typically building defects are classified as major or minor. Movement is a major defect, complex and expensive to rectify, whereas minor defects can often be resolved during normal maintenance activities. The Building Surveyor is able to advise owners of the extent and impact of all building defects in an existing building. They can advise on the technological, legal, environmental, social, economic and regulatory issues involved with the proposed retrofit project (RICS, 2020).

3.3.6 Covid-19 – transmission in high-rise HVAC systems

Covid-19, the virus which affected the world in 2020, was transmitted in some buildings through the ducting in HVAC systems. HVAC is common in commercial office buildings. The minute particles of the Covid-19 virus are transported in the air which is moved around by the HVAC systems, thereby enabling transmission from floor to floor. Concerns over the virus being transferred through contact, as well as airborne transmission, brought a shift and employees were quickly advised to work from home. The result was that city centres, Central Business Districts (CBDs) and Downtowns were deserted from the early part of 2020. At the time of writing, there has not been a full return to the office in most countries because of infection risks. This shows the enormous extent and vulnerability of the existing building stock to health-related risk that potentially renders buildings redundant. It remains to be seen in 2022 and beyond, how these risks are managed in respect of Covid-19 and other potential airborne diseases. Many services engineers will be investigating ways to isolate systems and parts of buildings from existing HVAC installations. Future design may be affected, which could lead to larger services ducts and less net lettable floor space being provided than previously or lower density development to adopt more natural ventilation. This may also be the case for residential and hotel buildings which have HVAC systems, not just commercial buildings.

3.4 The mismatch of buildings

Covid-19 changes to working patterns – working from home impact on property market – economic and environmental consequences. At the time of writing in 2022, the full economic and social impacts of Covid-19 are unknown, and there is speculation about future patterns of working. Willmot (2020) summarised the Australian office market situation in late December 2020 as:

> *The fund manager projects that office space demand will reduce by 700,000sq m over 2020 and 300,000sq m over 2021 from pre-Covid-19 levels.*
> *The big change is expected to come as demand for office space is hit by the accelerated trend towards work from home or more flexible arrangements.*
> *QIC says companies will hold on to their flagship central locations, and half the workforce is expected to work from home two days per week post-Covid-19.*
> *The fund manager says activity-based working will increase, moving from the traditional arrangement of one desk per employee to a more flexible structure with desk booking systems allowing for staff to reserve space.*
> *Investa, Head of Research and Strategy, David Cannington, says a significant number of tenants are unsure how they will use their workplaces and the top tier city landlord is in close contact with its blue-chip tenants about their intentions* (Willmot, 2020).

Initially, in 2020 and 2021, many city-based businesses reacted by requiring employees to work from home. As the rates of infection and transmission reduced or stabilised over time, some employees were able to return to office-based work for a part, or all, of the working week. Other organisations with multiple buildings reviewed their occupancy levels and determined to temporarily, and then later, permanently close some buildings. Over the course of 2020, some leases came to an end, and leasing agents reported lessees either not renewing leases or requesting reduced areas to lease. It is predicted that there could be migration of tenancies from lower grade stock to higher grade property as owners offer attractive lease deals to ensure higher rates of occupancy in that stock. There could be backfilling throughout the stock leaving the least attractive stock with high rates of vacancy. Alternate suitable uses will need to be found (Wilkinson et al., 2021a). This creates opportunities for revitalising city centres with a possible change of use adaptations (Wilkinson et al., 2021b). The issue will be that some change of use adaptations are hard to reverse and will lead to a permanent change of use in those locations. Others may be shorter term and non-permanent.

In retail, vacancy rates have increased as the sector transitioned to more online shopping. The smaller high street shops are particularly affected, as many are located on busy roads with no nearby car parking, which reduces passing foot traffic to a minimum. The result is higher vacancy rates across the stock and perceptions of inadequacy in respect of contemporary requirements of the stock.

This economic impact could lead to land uses previously not seen in city centres due to the high rents; there could be a resurgence in arts uses and more community activities as councils and local government seek to retain footfall and activity in these locations. Buildings deteriorate rapidly when unoccupied and are more vulnerable to vandalism and squatting, so continued use is desirable where possible. Table 3.1 summarises some of the potential issues affecting office, retail, residential and industrial stock and a SWOT analysis in respect of resilient retrofits.

There is a mismatch of buildings meeting the social needs of communities which has been exacerbated by the growth in inequality between the poorest and richest in society (Jaumotte et al., 2013). The figure that is there has been a 20% growth in inequality from 1980 to 2016 in the US (Pew Research Center, 2020), for example. This means more people have precarious employment and insecure accommodation, which manifests as increases in numbers of people begging on the streets and sleeping rough. Many sleep in their cars and are, in effect, the hidden homeless, masking the true extent of the problem. Ultimately, there is a high cost to society of poverty; it manifests in the court and health systems, in our jails and hospitals. Overall, it is more cost-effective to care for all in society to provide affordable housing, access to healthcare and employment opportunity. As noted in Table 3.1, higher vacancy rates in buildings may present opportunities for retrofit for more affordable and social housing and potential employment for the less fortunate in society.

With technological innovations there will be impacts on buildings, for example, future shifts in transport means, with autonomous vehicles, there may be less car ownership and with vehicles estimated to be used 22 hours a day compared to the current average daily use of less than 1 hour (Fagnant & Kockelman, 2015), less need for car parking. Porter et al. (2018) estimate 40% of vehicles will be autonomous by 2040.

Another issue, with a changing climate and/or increasing international disharmony, the issue of food security for those living in cities is becoming more pressing. Some advocate for the need to develop urban food production capacity to feed urban populations locally. Regardless of potential future conflicts impacting food supply chains, the potential to produce fresh food with low carbon food miles is very attractive. The post Covid-19 economic impact on supply and demand for existing buildings may create opportunities to convert buildings once in demand, such as redundant big box retail and commercial buildings.

Table 3.1 Summary Table of SWOT analysis for resilient retrofits for different property types

Sector	Strengths	Weaknesses	Opportunities	Threats
Office	Good range of buildings from small to large. Smaller buildings easier to adapt.	Larger buildings less flexible as alternate uses require different building regulations and standards.	Short-term leases to start-up companies and other non-traditional city centre businesses. Explore options for reversible alternate uses.	Overtime – loss of office stock in cities. One-way conversions – i.e. cannot return to office use.
Retail	Range of properties from small to large, range of owners from individuals to listed investment companies.	May be hard to find new users/alternate uses for property. Adaptive reuse may be expensive.	Short-term leases offered to start-up companies. Explore options for reversible alternate uses.	Lower office occupation leads to lower demand and permanent loss of retail space. Loss of local employment.
Residential	Large amount of individual investor-owned properties in some cities.	Lower demand may see rental levels fall below mortgage repayment levels. Owners decide to sell and move to regional cheaper areas.	Possibility for more social housing where homelessness is an issue.	Market collapses and potential for retrofit is very limited because of multiple ownership in higher rise apartment stock.
Industrial	Variety of spaces available.	Could be contamination from prior industrial uses. May not be possible to convert to other use if adjoining industrial uses ongoing.	Rezoning industrial use to new land use more suited to demand.	Loss of local employment if industrial land use lost.

(Source: Authors).

3.5 Conclusion

This chapter has outlined the complex and diverse challenges in relation to retrofitting the existing building stock. In summary, these challenges are environmental and economic, social and technological, regulatory and political. They can manifest independently and concurrently, and they can be acute or chronic, growing slowly over long periods or presenting sharply and unexpectedly as rapid onset shocks. For example, Covid-19 has pre-empted an acute unpredicted health shock, which may result in a chronic long-term economic impact. Managing the stock for optimum use and performance in these circumstances is challenging. The complexity is further exacerbated by the stakeholder groups who may include owners, investors, lessees, legislators and built environment professionals, all having different drivers and barriers to contend with.

Estimates within Europe are that fewer than 5% of buildings are 'energy efficient': far less are zero carbon. Given that legislation for minimum energy standards in Europe predates other countries by two decades and more, stock in other countries is even less energy efficient. Retrofitting for low energy is only part of the picture: water conservation, flood protection and fire resilience are increasingly important. Use of appropriate materials with low VOCs and low embodied energy is needed in retrofit projects. The challenge with sustainable, resilient retrofits is to balance what is retained and what is replaced and upgraded.

Caution is needed when upgrading buildings and applying new materials into older building systems without considering the building as a whole. Disasters, such as the Grenfell Tower fire, show this need for careful, holistic thinking, as the fire spread through the Tower's external envelope, having been retrofitted with new external cladding to address inadequacy in thermal performance. The rapid spread of the fire and occupants becoming trapped is evidence of the gaps in current legislation, design knowledge and understanding of the spread of fire and deadly smoke.

The health and well-being agenda is paramount, and research connects chronic and fatal conditions with pollution and building defects, so the quality of buildings being retrofitted needs evaluation by a Chartered Building Surveyor. The 2020 Covid-19 pandemic revealed a hitherto unknown risk of disease transmission through HVAC systems. This risk encouraged employers to get office workers to work remotely to reduce the risk of virus exposure and transmission. Working from home worked well for many staff and employers, and there are ongoing discussions about the extent of the return to work post-pandemic. This may lead to higher vacancy in the office sector. In which case, there will be more growth in our 'inadequate stock'. Collectively this creates a multifaceted challenge to adapt and improve our existing buildings.

Notes

1 See www.wellcertified.com/
2 See for example: Nolsoe (2020)
3 https://ec.europa.eu/energy/eu-buildings-factsheets-topics-tree/building-stock-characteristics_en

References

Avasarala, S., Wu, P. Y., Khan, S. Q., Yanlin, S., Van Scoyk, M., Bao, J., Di Lorenzo, A., David, O., Bedford, M. T., Gupta, V. and Winn, R. A., 2020. PRMT6 promotes lung tumor progression via the alternate activation of tumor-associated macrophages. *Molecular Cancer Research*, 18(1), pp. 166–178. https://doi.org/10.1158/1541-7786.MCR-19-0204.

Azari, R. and Abbasabadi, N., 2018. Embodied energy of buildings: A review of data, methods, challenges, and research trends. *Energy and Buildings*, 168, pp. 225–235. https://doi.org/10.1016/j.enbuild.2018.03.003.

Ball, M., 2003. Is there an office replacement cycle? *Journal of Property Research*, 20(2), pp. 173–189. https://doi.org/10.1080/0959991032000109535.

Bhattacharya Mis, N., Montz B., Proverbs, D., Chan, F., Kreibich, H., Lamond, J. and Wilkinson, S., 2018. *Advising on Flood Risk – Opportunities and Challenges Across International Commercial Property Markets*. RICS COBRA UCL, London.

Books, H. S. E., 2013. Managing and working with asbestos. In *Control of Asbestos Regulations 2012. Approved Code of Practice and Guidance*. HSE Books, Norwich.

Buckman, A. H., Mayfield, M. and Beck, S. B., 2014. What is a smart building? *Smart and Sustainable Built Environment*, 3(2), pp. 92–109. https://doi.org/10.1108/SASBE-01-2014-0003.

Bullen, P. A. and Love, P. E., 2011. Adaptive reuse of heritage buildings. *Structural Survey*. https://doi.org/10.1108/02630801111182439.

Corna, A., Fontana, L., Nacci, A. A. and Sciuto, D., 9–13 March 2015. Occupancy detection via iBeacon on Android devices for smart building management. In *2015 Design, Automation & Test in Europe Conference & Exhibition* (pp. 629–632). IEEE, Grenoble, France. https://doi.org/10.7873/DATE.2015.0753.

Dean, B., Dulac, J., Petrichenko, K. and Graham, P., 2017. *Global Status Report 2016: Towards Zero-Emission Efficient and Resilient Buildings*. Global Alliance for Buildings and Construction (GABC), Paris.

Dobson, D. W., Sourani, A., Sertyesilisik, B. and Tunstall, A., 2013. Sustainable construction: Analysis of its costs and benefits. *American Journal of Civil Engineering and Architecture*, 1(2), pp. 32–38. https://doi.org/10.12691/ajcea-1-2-2.

Ehrenreich, J., 2016 *Third Wave Capitalism: How Money, Power, and the Pursuit of Self-Interest Have Imperiled the American Dream*. Cornell University Press, Ithaca, NY.

Environmental, U. S., February 1991. Indoor Air Facts No. 4 Sick Building Syndrome. *EPA-Air Radiation(6609J), Research and Development*, pp. 1–4.

Evanoff, B. A., Strickland, J. R., Dale, A. M., Hayibor, L., Page, E., Duncan, J. G., Kannampallil, T. and Gray, D. L., 2020. Work-related and personal factors associated with mental well-being during the COVID-19 response: Survey of health care and other workers. *Journal of Medical Internet Research*, 22(8), p. e21366. https://doi.org/10.2196/21366.

Fagnant, D. J. and Kockelman, K., 2015. Preparing a nation for autonomous vehicles: Opportunities, barriers and policy recommendations. *Transportation Research Part A: Policy and Practice*, 77, pp. 167–181. https://doi.org/10.1016/j.tra.2015.04.003.

Fortino, G., Guerrieri, A., O'Hare, G. M. and Ruzzelli, A., 2012. A flexible building management framework based on wireless sensor and actuator networks. *Journal of Network and Computer Applications*, 35(6), pp. 1934–1952. https://doi.org/10.1016/j.jnca.2012.07.016.

Gov UK. 2020. *English Housing Survey 2019 to 2020: Headline report*. Ministry of Housing, Communities & Local Government. Available from www.gov.uk/government/statistics/english-housing-survey-2019-to-2020-headline-report. Accessed 8 January 2022.

Grenfell Tower Inquiry Report, 2019. Available from www.grenfelltowerinquiry.org.uk/phase-1-report. ISBN 978-1-5286-1602-7. Accessed 27 December 2020.

Harris, R., 2020. *The Age of Unreal Estate: The Changing Use and Value of Commercial Property*. Royal Institution of Chartered Surveyors. Available from www.rics.org/globalassets/rics-website/media/knowledge/research/insights/the-age-of-unreal-estate_1st_edition.pdf. Accessed 13 February 2022.

Jaumotte, F., Lall, S. and Papageorgiou, C., 2013. Rising income inequality: Technology, or trade and financial globalization? *IMF Economic Review*, 61(2), pp. 271–309. https://doi. org/10.1057/imfer.2013.7.

Keywood, M. D., Emmerson, K. M. and Hibberd, M. F., 2016. Ambient air quality: Volatile organic compounds. In *Australia State of the Environment 2016, Australian Government Department of the Environment and Energy*. Canberra. DOI 10.4226/94/58b65c70bc372. Available from https://soe.environment.gov.au/theme/ambient-air-quality/topic/2016/ volatile-organic-compounds. Accessed 13 February 2022.

Kirby, A. and Kent, T., 2010. The local icon: Reuse of buildings in place marketing. *Journal of Town & City Management*, 1(1).

Lamond, J. E., Bhattacharya-Mis, N., Chan, F. K. S., Kreibich, H., Montz, B., Proverbs, D. G. and Wilkinson, S., 2019. Flood risk insurance, mitigation and commercial property valuation. *Property Management*.

Lindsey, R., 14 August 2020. *Climate Change: Global Sea Level*. NOAA. Available from www.climate.gov/news-features/understanding-climate/climate-change-global-sea-level. Accessed 27 Dec 2020.

Mansfield, J. R. and Pinder, J. A., 2008. "Economic" and "functional" obsolescence: Their characteristics and impacts on valuation practice. *Property Management*, 26(3), pp. 191–206. https://doi.org/10.1108/02637470810879233.

Medal, L. and Kim, A., 2020, November. Context-driven factors for implementing energy efficiency retrofit in a portfolio of buildings. In *Construction Research Congress 2020: Infrastructure Systems and Sustainability* (pp. 491–500). American Society of Civil Engineers, Reston, VA.

Mekonnen, M. M. and Hoekstra, A. Y., 2016. Four billion people facing severe water scarcity. *Science Advances*, 2(2), p. e1500323.

Nolsoe, E., 2020. Pandemic has made properties with gardens more attractive. Available from https://yougov.co.uk/topics/finance/articles-reports/2020/08/06/pandemic-has-made-properties-gardens-more-attracti. Accessed 13 February 2022.

Palumbo, R., 2020. Let me go to the office! An investigation into the side effects of working from home on work-life balance. *International Journal of Public Sector Management*, 33(6/7), pp. 771–790. https://doi.org/10.1108/IJPSM-06-2020-0150.

Pascale, S., Kapnick, S. B., Delworth, T. L. and Cooke, W. F., 2020. Increasing risk of another Cape Town "Day Zero" drought in the 21st century. *Proceedings of the National Academy of Sciences*, 117(47), pp. 29495–29503.

Passarelli, G. R., 2009. Sick building syndrome: An overview to raise awareness. *Journal of Building Appraisal*, 5(1), pp. 55–66. https://doi.org/10.1057/jba.2009.20.

Pew Research Center, 2020. Trends in income and wealth inequality. Available from www. pewsocialtrends.org/2020/01/09/trends-in-income-and-wealth-inequality/. Accessed 28 December 2020.

Porter, L., Stone, J., Legacy, C., Curtis, C., Harris, J., Fishman, E., Kent, J., Marsden, G., Reardon, L. and Stilgoe, J., 2018. The autonomous vehicle revolution: Implications for planning/the driverless city? Autonomous vehicles – a planner's response/autonomous vehicles: Opportunities, challenges and the need for government action/three signs autonomous vehicles will not lead to less car ownership and less car use in car dependent cities – a case study of Sydney, Australia/planning for autonomous vehicles? Questions of purpose, place and pace/ensuring good governance: The role of planners in the development of autonomous vehicles/putting technology in its place. *Planning Theory & Practice*, 19(5), pp. 753–778. https://doi.org/10.1080/14649357.2018.1537599.

Roberts, S., 2008. Altering existing buildings in the UK. *Energy Policy*, 36(12), pp. 4482–4486. https://doi.org/10.1016/j.enpol.2008.09.023.

Royal Institution of Chartered Surveyors (RICS), 2020. *Technical Due Diligence Guide of Commercial Property*. 1st edition. ISBN 978 1 78321 378 8.

Sayce, S., Ellison, L. and Smith, J., 2004. Incorporating sustainability in commercial property appraisal: Evidence from the UK. *Australian Property Journal*, 38(3), pp. 226–233.

Scott, A. J., 2019. Land redevelopment and the built environment in third-wave cities: Review and synthesis. *Journal of Urban Technology*, 26(1), pp. 57–81. https://doi.org/10.1080/10630732.2018.1537050.

Straus, D. C., 2009. Molds, mycotoxins, and sick building syndrome. *Toxicology and Industrial Health*, 25(9–10), pp. 617–635. https://doi.org/10.1177/0748233709348287.

Terr, A. I., 2009. Sick building syndrome: Is mould the cause? *Medical Mycology*, 47(Supplement_1), pp. S217–S222. https://doi.org/10.1080/13693780802510216.

Wilkinson, S., 2011. *Sustainable Retrofit Potential in Lower Quality Office Stock in the Central Business District*. CIB Management and Innovation in the Sustainable Built Environment. Delft University of Technology, Delft, Amsterdam. ISBN: 9789052693958.

Wilkinson, S., Armstrong, G. and Cilliers, J., 2021a. We're gonna make you a STAR. Sustainable Temporary Adaptive Reuse in the CBD. AIQS December. Available from https://protect-au.mimecast.com/s/ToftCOMKZohGLy4qTEfUug?domain=issuu.com. Accessed 13 February 2022.

Wilkinson, S., Armstrong, G. and Cilliers, J., December 2021b. Repurposing assets to reduce vacancy rates. *RICS Property Journal*. Available from https://ww3.rics.org/uk/en/journals/property-journal/repurposing-assets-to-reduce-vacancy-rates.html. Accessed 13 February 2022.

Wilkinson, S. and Kalejska-Jonsson, A., 2019. The Relationship Between Building Performance and Human Behaviour, Conference Paper, 14–16 January, PRRES University of Melbourne, Australia.

Wilkinson, S. J., Van Der Kallen, P. and Leong Phui, K., 2013. The relationship between occupation of green buildings, and pro-environmental behaviour and beliefs. *The Journal for Sustainable Real Estate*, 5(1), pp. 1–22. https://doi.org/10.1080/10835547.2014.1209 1850.

Wilkinson, S. J. and Zajelska-Jonsson, A., 2021. Student accommodation, environmental behaviour and lessons for property managers. *Property Management*. https://doi.org/10.1108/PM-09-2020-0055.

Willmot, B., 2020. Questions arise over workers returning to the office in 2021. Available from www.realcommercial.com.au/news/coronavirus-workers-choice-will-you-or-wont-you-return-to-the-office?rsf=or:twitter. Accessed 28 December 2020.

World Health Organization, 2010. *WHO Guidelines for Indoor Air Quality: Selected Pollutants*. World Health Organization. Regional Office for Europe, Copenhagen.

Zhang, D., Shah, N. and Papageorgiou, L. G., 2013. Efficient energy consumption and operation management in a smart building with microgrid. *Energy Conversion and Management*, 74, pp. 209–222.

4 Understanding vacancy in the office stock

Gillian Armstrong

4.1 Introduction

Given the uncertain impacts of Covid-19 stresses on cities, the need for greater vacancy understanding is particularly urgent. Current concerns predict high levels of commercial building vacancy, and there is an awareness of an increased risk of premature obsolescence (Fenton, 2021). This chapter aims to make a case for a more nuanced understanding of vacancy as a valuable evidence base for mitigating obsolescence and building urban resilience. The need for this evidence base has emerged from surprising data findings exploring existing building adaptation in urban centres. The results from vacancy data challenge the accepted wisdom of one popular strategy to build urban resilience by converting vacant office buildings to new uses, referred to as 'adaptive reuse' or conversion. This chapter also offers suggestions on how to advance vacancy knowledge. It discusses insights from developing a practical tool for policymakers to quantify vacancy, known as Vacancy Visual Analytic Method (VVAM) (Armstrong et al., 2021). Finally, this chapter highlights the usefulness of vacancy as an essential evaluation tool in policy development to address chronic stresses and acute shocks experienced by cities.

Vacancy is a precursor to premature obsolescence in the built environment (Muldoon-Smith, 2016; Remøy & Street, 2018). Changes in how we value and use existing buildings can create vacancy or underused space within parts of buildings. Over time, these changes can result in a building becoming more and more under-occupied and at risk of obsolescence. The climate emergency demands better management of underused buildings at risk of premature obsolescence (Wilkinson & Remøy, 2018). The causes of built environment obsolescence are varied. However, obsolescence essentially involves a lack of local demand for a building as it currently stands and a lack of active decision-making to address underuse and vacancy (Aigwi et al., 2020). A range of active adaption strategies to mitigate the risk of premature obsolescence sit at the single building scale (Greenhalgh & Muldoon-Smith, 2017). However, vacancy is multifaceted and an intrinsic part of any complex and dynamic city (Buitelaar et al., 2021). The effects of vacancy can reach beyond the physical boundary of a single obsolete building, impacting urban centres on a scale much greater than the building itself.

DOI: 10.1201/9781003023975-5

One crucial element in developing an effective urban policy to address pre-mature obsolescence is a robust and critical understanding of the vacancy itself. Without a sound conceptual understanding of the vacancy and fine-grain mechanisms to document and measure vacancy, policy development is working blind to develop the capacity to avoid premature obsolescence and build urban resilience.

This chapter first sets out the background context to vacancy before presenting VVAM, a practical tool for policymakers to quantify and visualise vacancy. Finally, the chapter presents data findings from applying VVAM to two building populations perceived to have high vacancy levels and an increased risk of obsolescence.

4.1.1 Vacancy development by professions and in policy

Policy action to address underuse or vacancy has its roots in heritage protection advocacy work in the 1960s and 1970s. In Italy, The Venice Charter first set out principles of managing change by proposing new uses for at-risk vacant heritage buildings. Soon after, translations of these mechanisms or jurisdictions beyond Italy were developed, including Australia's The Burra Charter (Saniga, 2012). This heritage policy set out the principles for better managing vacancy and keeping heritage buildings in use despite cultural, economic and technical shifts experienced in urban and rural places.

Vacancy rates have a long history in assisting the management of buildings (Bodfish, 1931). Facilities management (FM) is a growing specialised profession that recognises the need to mitigate changes in existing buildings, including strategies by complex organisations to manage occupancy and vacancy in stocks of space (Best et al., 2003). Together, the established heritage policy and emerging focus on change in existing buildings by FM professionals and others link the need to address vacancy across a full spectrum of existing building stocks – both old and non-heritage buildings.

Cities can both shrink and grow, often simultaneously in complex ways. Vacant space is a crucial indicator of shrinkage and growth, and data provides insightful perspectives for urban planners (Wolff & Wiechmann, 2014). Vacancy information can aid local government plans by guiding future development. Vacancy information can aid local government plans by guiding future development. However, the recognition of vacancy data for managing new urban development is yet to be realised by strategic and town planners (Burkholder, 2012). Further, a lack of vacancy data can lead to the displacement of communities and employment through gentrification and where buildings are not fully vacant or whose uses are considered unviable or redundant on economic grounds (Grodach et al., 2017).

4.1.2 Scenarios of vacancy from around the world

Manifestation of vacancy appears whenever countries, cities and downtown areas experience acute stresses and shocks. No building or location is immune where vacancy manifests as high average vacancy rates (%) across specific building stocks, for example, commercial buildings, industrial infrastructure or residential markets. Next, this chapter presents examples of vacancy scenarios and highlights the causes and impacts vacancy can inflict on cities. The scenarios presented examine vacancy at the scale of a district or precinct.

Economic changes from industrial shifts are well documented in the literature, depicting wide-scale vacancy and long-term urban decay, often in fringe areas or smaller towns. Solutions to economic changes focus on reinvention, developing new uses for empty buildings and sites on a large scale in fringe locations. Recent attention focuses on the impact of vacancy in The Rust Belt, US, in cities such as Detroit and Michigan. Sustainable solutions to the vacancy in The Rust Belt are varied. Solutions include reshaping buildings and vacant sites by new waves of immigrant populations and new public-private investment to develop hi-tech industries such as robotics. Similarly, regeneration can happen through developing knowledge services by universities and healthcare services (Pottie-Sherman, 2020). Elsewhere, attention to similar industrial shifts in Italy poses solutions to vacancy through conversion of vacant industry buildings into hotel accommodation for tourists (Bottero et al., 2019) or where the industrial heritage aesthetics are woven into new touristic experiences and are valued as part of new 'consumption-scapes' such as restaurants or craft breweries, a trend in many cities worldwide (Lynch, 2021).

For vacant space in commercial business districts and city centre fringes, solutions often suggest transitioning empty commercial buildings to new uses, particularly residential apartments (Remøy, 2010). However, there are several strategies available other than adaptive reuse (Greenhalgh & Muldoon-Smith, 2017). Indeed, voices are increasingly critical of adaptive reuse as a final destination for commercial office buildings. Low-quality adaptive reuse can occur if technical solutions do not meet minimum regulation codes, and adaptive reuse development falls short of planning and performance standards expected of new build development (Clifford et al., 2018).

Technological shifts resulting in high vacancy are emerging in retail building stocks. As the shift to online shopping occurs globally, relatively recent shopping malls and 'big-box' retail buildings face premature obsolescence (Roberts & Carter, 2020; Lesneski, 2011). Impacts of the Covid-19 pandemic on space in retail buildings have accelerated this shift to online shopping further at present (Savills, 2020). The transition of these vacant buildings is currently experimental. Suggestions for addressing vacancy on such large-scale retail complexes look towards adaptive reuse for large-scale aged-care or one-stop-shop purposes with large floor plate requirements, bringing together multiple agencies working.

Two decades of social and political change (1980s–1990s) during the breakdown of South Africa's apartheid regime brought a series of unexpected and quite sudden events that generated vacant space in the four major city centres of Johannesburg, Cape Town, Pretoria, and Durban (Turok et al., 2019). The political and social transitions brought an exodus from city centres to outer suburbs by many white property owners, investors and occupiers (Winkler, 2013). Solutions to the widespread vacancy have emerged via community activism and partnerships between enterprising developers, stakeholder groups and metropolitan municipalities. From a sustainability perspective, the existing structures have not changed much as the buildings themselves have proved robust and slow to change. Land-use diversification has occurred within the retained building shells, moving away from a more sterile traditional commercial business district (CBD). The solutions to high vacancy are still uneven 30 years later. This unevenness highlights the complexity of occupancy and vacancy in existing buildings, with vibrant solutions and urban decay side-by-side in all four cities following the sudden but necessary social and political change (Turok et al., 2019).

The social restrictions to control the spread of Covid-19 has resulted in a similar migration from city centres to suburbs (Felstead & Reuschke, 2020; Fenton, 2020; Rosenthal & Rothenberg, 2021). Sharp rises in vacancy in commercial buildings were experienced in 2020 in cities, and up until 2020, cities have suffered from too little vacancy, such as London, New York and Sydney. It is too early in the pandemic to predict whether the sudden pandemic event will produce high long-term structural vacancy rates. Recovery from Covid-19 will need to address any redistribution of occupancy and vacancy.

Categories of obsolescence developed by Barras and Clark's (1996) early work can be seen in the scenario examples given earlier (Barras, 2009). These categories include broader economic and social or political changes, physical and functional limits of a building and legal changes often resulting from the risk of loss of life. Similarly, the scenario examples highlight sustainable solutions to vacancy, which need to be critically considered through a resilience lens, depending on the scale and distribution of different vacancy types.

4.2 Current vacancy understanding

There are few theoretical models to understand vacancy, and current thinking tends to rely upon Atkinson's (1988) 'sinking stack' theory for housing stocks. The sinking stack theory proposes that in older buildings, vacancy increases as new buildings are completed and enter the market (Abdullah et al., 2020). This theory suggests buildings 'sink' through the quality grades until their final disposal when they reach peak vacancy and are considered obsolete. Ness (2002) proposes vacancy can sink through building stock quality grades as an 'indigestible lump' when there is an oversupply of new commercial buildings. A further assumption suggests that tenants move to higher quality buildings when vacancy rates are high if rents in more contemporary structures are comparable (Remøy & van der Voordt, 2014). However, no research studies have tested the sinking stack theory, nor for different building stocks or property markets over time. There is little testing of the theory to examine if it holds true in all geographic locations or property markets at different points in every economic cycle (Armstrong, 2020).

A philosophy of 'minimum decay' is helpful to justify the economic cost of adaptions, where adaptions can address vacancy and slow the rate of obsolescence (Atkinson, 1988). This philosophy involves planning for the latter stages of a building's life cycle by 'assigning increased resources to maintenance and refurbishment' (Ness & Atkinson, 2001:2). Sustainable actions to slow the rate of obsolescence and tackle vacancy early are well documented and include consolidation of vacant space, corrective maintenance, extensions and 'top-up' additions, creative demolition, retrofit and energy efficiency upgrades, 'meanwhile' and alternative reuses, property sale and deconstruction (Holden, 2018; Wilkinson, 2018; Greenhalgh & Muldoon-Smith, 2017). These actions align vacancy amelioration with circular economy principles of reduce, reuse, recycle (Foster, 2020).

The insufficient attention to vacant space in research is currently problematic as vacancy rates are often cited in public debate to be a driver of calls for policy action. Yet vacancy data are not routinely collected independently from commercial groups (Armstrong et al., 2021). Vacancy data collected by commercial property groups often do not publish details of how the data are collected, sample sizes, nor which buildings are included or excluded from the data calculations (Armstrong, 2020). Potential commercial biases are problematic when vacancy statistics are relied upon for policy development in cities. Furthermore, average percentage rates across large building stocks can only offer a simplistic understanding of vacancy. A lack of fine-grain data to understand where vacancy is sitting also prevents meaningful explorations for policy development, resulting in ad hoc reactive policy to address vacancy and its impacts (Buitelaar et al., 2021).

The reasons briefly highlighted earlier describe some of our significant gaps in understanding vacant space at present. These gaps call for urgent attention by policymakers and researchers to value and collect vacancy data. Attention to these gaps is critical if we are to develop to mitigate obsolescence at an earlier stage of the building life cycle so that problems in building stocks can be detected and critically explored much earlier. At a city or precinct scale, the vacancy distribution can then be evaluated and used to inform policy action to aid solutions and build capacity for greater urban resilience.

4.2.1 Vacancy types

This section sets out the complexity in vacancy understanding, the underdeveloped concept of unused space. Attempts to develop a conceptual framework of vacancy are emerging from Muldoon-Smith and Greenhalgh (2017) through an exploratory conference paper, which describes the vacancy process across the commercial building life cycles and property markets. This framework provides a deeper discussion of vacancy types, developing an understanding of vacancy from rather simplistic binary notions of natural vacancy, from which buildings recover, and structural vacancy, which speeds up obsolescence. The remainder of this section pulls together the published research which currently explores vacancy as a concept relating to non-residential buildings. Table 4.1 summarises vacancy concepts discussed in this chapter relating to adaptation of non-residential buildings, signalling where these different concepts or definitions have originated from and the defining characteristics of each. Table 4.1 is by no means exhaustive but provides a gateway into this understudied field of vacancy, which can be helpful as an early indicator of premature obsolescence in non-residential buildings and

Table 4.1 Types of vacancy for non-residential buildings

Vacancy type	Key characteristics	Originating research publications
Untenanted vacancy	• space which is advertised for lease	Armstrong (2020)
Premium vacancy	• exists in the very best quality buildings with high standards of design, technology and amenities • temporary – due to tenants relocating or newness of construction	Muldoon-Smith and Greenhalgh (2017)
Auxiliary vacancy	• exists in quality secondary grade buildings • temporary – when low vacancy in premium buildings • longer term – when a mismatch between supply of premium grade space and low demand for space	Muldoon-Smith and Greenhalgh (2017)
Structural vacancy	• typically exists in where low demand for space or demand for use • left empty for more than three years and deemed untenable	Remøy, 2010
Evolutional vacancy	• a sub-category of structural vacancy • can be resolved through maintenance and amenity and/or performance upgrade	Muldoon-Smith and Greenhalgh (2017)
Final vacancy	• a sub-category of structural vacancy • can only be resolved through change of use, due to local property market conditions	Muldoon-Smith and Greenhalgh (2017)
Greyspace vacancy	• space which is surplus to existing occupants requirements	Muldoon-Smith and Greenhalgh (2017) Muldoon-Smith (2016) Armstrong et al. (2021)

(Source: Armstrong & Wilkinson).

can inform sustainable asset management at a building level (Armstrong, 2020; Muldoon-Smith, 2016; Remøy, 2010) and urban policy decisions at a city level (Armstrong, 2021; Buitelaar et al., 2021). The authors acknowledge that more research is needed to unpack these terms further and undertake applied studies to fully understand the role of vacancy in building obsolescence.

Muldoon-Smith and Greenhalgh (2017) propose that 'premium vacancy' is temporary due to cyclical markets, caused by frictional movement of tenants relocating or initial vacancy in any new development. They define premium vacancy as unlet space which is sitting in 'the very best buildings that are on the market' (p. 482). The authors also suggest 'auxiliary vacancy' exists only temporarily in high-quality secondary grade properties if prime buildings are in short supply and demand is high (p. 482). Though premium and auxiliary vacancy are explained as two forms of temporary vacancy, a caveat explains that auxiliary vacancy may be more permanent when property markets have stagnated with adverse economic conditions. Structural vacancy is conceived to be vacancy which is terminal, left empty for more than three years and is deemed to be untenantable (Muldoon-Smith, 2016; Remøy, 2010). Structural vacancy can comprise a mixture of 'evolutionary vacancy' and 'final vacancy'. As the name suggests, evolutionary vacancy can be resolved through adaption, for example maintenance, adaptation upgrades through changing a building's interior design, building performance. However, final vacancy is much harder to remedy as the building is at significant risk of obsolescence unless it undergoes a change of use. It is often suggested that properties with high levels of final vacancy should transition out of their current property market altogether. However, this conceptual framework bases itself on the sinking stack theory, which needs more critical exploration by research. One further, and important, type of vacancy, Greyspace vacancy, is described as a 'hidden' vacancy type as it is concealed in space already leased, but it is surplus to tenants' requirements and is therefore underused. Before VVAM, this vacancy type was challenging to locate and quantify (Muldoon-Smith & Greenhalgh, 2017).

The phenomenon of greyspace vacancy can affect large floor areas within buildings and mask the true extent of obsolescence with a building or across a market as it is not counted as untenanted vacancy despite being unused (Muldoon-Smith, 2016:115; Hammond, 2013). As clarified by VVAM, quantification of greyspace is possible through data that capture floor use by tenants, such as mandatory energy rating certification programmes and data collected by local government for calculating annual rates for non-residential properties.

Commercial organisations publish vacancy analysis reports. However, without access to the data, it is not possible to undertake deeper categorisation of vacancy types helpful for policy decision-making. Published vacancy data are only presented as 'untenanted vacancy' which is simply space that is advertised as 'for lease' and quantified by real estate market agents (Armstrong, 2020).

Yet different vacancy types can have significant impacts on cities but rarely appear in data analysis and vacancy discussions. One explanation for this is that there is little consensus between different vacancy types, and the language to describe vacancy processes is not yet developed (Muldoon-Smith & Greenhalgh, 2017). Understanding and types are emerging, however. For example, there is a consensus on the definition of 'structural vacancy' which connects closely with the final stages of obsolescence of a building and decisions to invest and adapt. However, structural vacancy needs longitudinal data analysis as it is only detectable over time.

A further challenge is that non-commercial organisations do not regularly collect vacancy data, and as such, fine-grain longitudinal studies are unlikely to be achievable at present. Undertaking cross-sectional studies are more realistic at present, given the current lack of attention to analyse vacancy data in practice and research. In recognition of this challenge, it is pragmatically highlighted that untenanted vacant space is one type of vacancy detectable in cross-sectional data analysis. In addition, alongside untenanted vacancy, greyspace is suitable for analysis in cross-sectional studies (Armstrong et al., 2021).

In summary, a nuanced understanding of vacancy is still in its infancy, partly hindered by poor access to data, a lack of interest by non-commercial organisations to routinely collect the data and a relative lack of research to develop conceptual understanding to shape urban policymaking.

4.2.2 Data sources for using VVAM

Although vacancy data is believed to be hard to access, there are some recognised sources of data capable of quantifying vacancy in existing buildings. Data sources already being used as proxy measures for vacant space in different building stocks in Australia include:

- Data from energy efficiency certification schemes, such as Australia's NABERS commercial buildings rating scheme (Armstrong et al., 2021).
- Water usage meter data (Fitzgerald, 2020).
- Australian Bureau of Statistics five-yearly census data (Fitzgerald, 2020).
- Local government property valuation (Armstrong et al., 2021).
- Local government employment and floor space use five-yearly survey (City of Sydney, 2017).

A method to use secondary data to quantify, analyse and visualise vacancy in different building stocks is published (Armstrong et al., 2021). This method is described next in this chapter. The challenge now is to recognise the need for fine-grain analysis of vacancy data as a powerful analysis to inform policy development for urban resilience.

4.3 VVAM explained

Vacancy Visual Analytics Method (VVAM) is a practical tool developed to repurpose secondary data to understand vacancy and its distribution in existing building stocks. The shape and location of vacancy are important but cannot be deduced from aggregated vacancy rates (%) for entire building stocks. The quantitative method provides valuable data to build an understanding of vacancy at a local level. This understanding is essential to critically interrogate, or triangulate, findings from qualitative studies, such as findings from stakeholder interviews. Critical awareness is needed to reduce potential bias in descriptive reporting of stakeholder views of the causes of vacancy and invested interests in continuing unsustainable practices.

The need to develop a method to quantify vacancy came about via a study exploring barriers to greater uptake of adaptive reuse in underused office buildings. The study examined vacancies in a city centre perceived to have high vacancy levels, typical of many city centres suffering economic decline (Armstrong, 2020). The study's findings challenged in-grained perceptions about the 'go-to' solutions and policy mechanisms routinely assumed to be the solution for ameliorating vacancy and premature obsolescence in the built environment (Armstrong et al., 2021). VVAM has since been applied to a population of heritage buildings considered at risk of obsolescence and demolition.

VVAM aligns with longstanding calls from policymakers for research tools to be practical and realistic to apply (Lavis et al., 2004). VVAM can be applied to any building population for which space use data are collected. Figure 4.1 breaks down VVAM into its constituent parts for others to reproduce, including preparing secondary data, building an appropriate sample,

Quantifying and visualising vacancy data, before using this new vacancy understanding to inform policy development to build resilience in cities. VVAM makes a unique contribution for local government joining the drive to adopt a data-informed policy process in Australia (ACELG, 2013; CoM, 2013; Petit et al., 2020) and internationally (Goldsmith & Crawford, 2014).

4.3.1 Overview of the method

While secondary analysis is an efficient, flexible and valuable approach, it still requires a systematic process for any research findings to be robust. The method adapts Johnston's (2017) framework for repurposing data to aid in recognising and acknowledging any challenges and limitations of using secondary data. For example, when third parties collect data for other purposes, there can be inherent issues to resolve or acknowledge as limitations. Johnston's framework also offers a practical step-by-step approach to guide repurposing data for new reuses, such as VVAM. The systematic approach by Johnston also aids the reproducibility of studies that rely on secondary data so that the method can be further tested, carefully adjusted for new specific inquiries and local contexts.

A crucial step from Johnston (2017 is to evaluate the variables in the data to determine whether vacancy can be quantified and what types of vacancy can be deduced. This step needs to be done at the earliest time possible to avoid wasting efforts and is part of the secondary data preparation shown in Figure 4.1. First, the dataset is evaluated according to whether it is possible to identify the component of a building occupied and whether this occupation can be deducted from the overall gross lettable floor area in a building. For further reading on VVAM's steps to prepare data for interrogating vacancies, see Armstrong et al., 2021.

A building population sample is then determined using careful application of criteria for inclusion. A representative sample of buildings or the largest possible

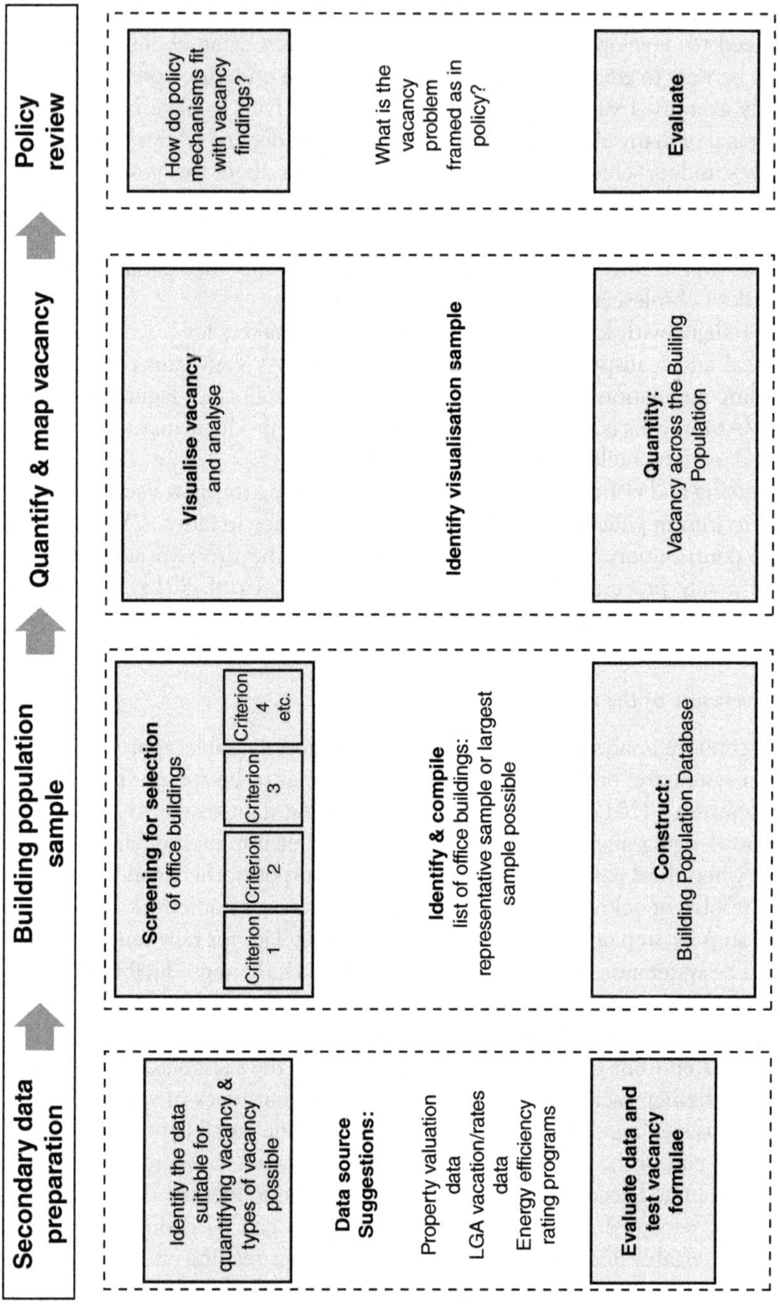

Figure 4.1 Overview of VVAM

(Source: Author)

selection can be established, depending on the available data. A database of buildings in the sample can be built. It can include supplementary sources of information relevant to the research, such as whether a building is listed, the date of construction of the building or the mix of functional uses within each building. The criteria for the selection of buildings should be led by the specific research inquiry or research aims. For instance, if decarbonisation through adaption of existing buildings is a primary objective, the sample may include all large-scale, multi-storey commercial buildings over ten years old. Or suppose the primary objective is to identify underused commercial buildings to address vibrancy and economic downturns. In that case, the sample may include only sites in prime locations as catalysts for change.

For heritage buildings at risk of further decay, the selection may consist of buildings on local and state heritage registers, and unlisted buildings identified as important in placemaking. For precinct renewal, understanding vacancy is also important to minimise displacement, as buildings are often not wholly vacant and existing occupants need to be considered.

VVAM uses stacked bar charts to represent vacancy and occupancy within each building and visualise the 'shape' and approximate location of different vacant and occupied spaces within a single structure. Figure 4.2 shows an example of the visualisations from VVAM. Stacked bar charts represent the data in a similar way to architectural section drawings. Sectional drawings differ from horizontal plans as they are used to communicate vertical dimensional information and, therefore, visualise occupancy and vacancy on a level-by-level basis. Microsoft Excel is used to create stacked bar charts because MS Excel software licences are typically standard software. Excel is also often used by local government and real-estate industry groups for collating property data and is, therefore, a practical choice.

Gross lettable area data (GLA, m²) is used to quantify occupancy and vacancy in VVAM. The method assumed GLA data to be calculated using the International Property Measurement Standards (API, 2017a, 2017b). It is important to note that GLA data cannot represent the actual floor area in any building. International measurement practice specifies what space is included and excluded from GLA calculations. These exclusions are dependent on the architectural design and location of services, including lifts and common stairs (API, 2017a, 2017b). Therefore, there can be some variation in the representation of a building's total floor area compared to the expected total floor area on a floor-by-floor basis.

For simplicity, the example in Figure 4.2 shows a building exclusively devoted to office uses. The distribution of occupancy and vacancy types shown in Figure 4.2 are as follows: occupied space in dark blue, office greyspace vacancy (oVG) coloured grey and areas of untenanted vacancy referred to by Armstrong (2020; Armstrong et al., 2021) as office Gross Lettable Area Untenanted (oGLAU), shown in light blue. Existing buildings often contain more than one functional use, for example, commercial buildings can include a combination of office-use, education or training-use, childcare provision and retail or restaurants. Therefore, the stacked bar charts can also visualise data across a range of functional uses (office, retail or other). In Australia and the UK, the classifications of building uses are set

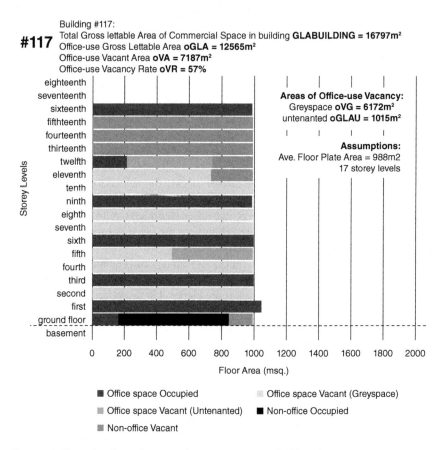

Figure 4.2 Example of visualisation of vacancy using stacked bar charts
(Source: Author)

by each countries building regulations. VVAM acknowledges that buildings are usually made up of separate suites of owners and tenancies.

The formula used in VVAM is shown in Figure 4.3. The first step calculates the Total Gross Lettable Area for each building (see Figure 4.3, formula 1). Preparatory work to accommodate the co-existence of different space uses involves applying codes or prefixes to the data. For example, to examine only office-use space in a building containing a mixture of uses, the prefix 'o' for office space use or 'r' for retail space use identifies different space uses (see Figure 4.3, formula 2). Applying different space uses enables GLAs to be calculated for a range of other building populations. VVAM can also accommodate mixed-use development, allowing VVAM to be flexible in developing effective policy action to address vacancy in real-world complex building settings.

The following steps or formula are used for each building using the formula on a building-by-building basis:

Formula 01 *To calculate a building's floor area:*

$GLA_{BUILDING} = GLA_1 + GLA_2 + GLA_3 + \cdots\cdots$ [m^2]

GLA_1, GLA_2, GLA_3, *etc. are the separate suites in a single building.*

Formula 02 *To calculate floor areas by class use:*

$GLA_{BUILDING} = \text{Total } oGLA_1 + \text{Total } rGLA + \cdots\cdots$ [m^2]

$\text{Total } oGLA = oGLA_1 + oGLA_2 + oGLA_2 + \cdots\cdots$ [m^2]

oGLA is the office-use space

$\text{Total } rGLA = rGLA_1 + rGLA_2 + rGLA_3 + \cdots\cdots$ [m^2]

rGLA is the retail space

Formula 3 *To calculate vacant area*

$VA = \text{Total } GLA_{BUILDING} - \text{Total } CGLA_{BUILDING}$ [m^2]

Formula 5 *To calculate untenanted and greyspace area:*

If $CGLA = 0$ m^2, untenanted vacancy $GLAU_{BUILDING} = \Sigma GLA_1 + GLA_2 + \cdots\cdots$ [m^2]

If $CGLA > 0$ m^2, greyspace vacancy $VG_{BUILDING} = \Sigma (GLA_1 - CGLA_1) +$

$(GLA_2 - CGLA_2) + \cdots\cdots$ [m^2]

CGLA is the space use declared by the end-users or building owners in the dataset

Formula 6 *To calculate occupancy and vacancy rates*

$OR = \text{Total } CGLA_{BUILDING}/\text{Total } GLA_{BUILDING} \times 100\%$ [%]

$VR = 100\% - OR$ [%]

Figure 4.3 Formula used in VVAM

(Source: Author)

Once the GLA area is calculated, vacant area can be quantified by identifying the difference between the GLA and GLA component (CGLA) occupied (see Figure 4.3, formula 3). Vacancy can be disaggregated further to enable a more nuanced and valuable understanding of space usage and demand within a city. Data from space advertised 'for lease' are only one component of urban vacancy.

To identify Untenanted (GLAU) vacancy and Greyspace vacancy (VG), datasets need to be collected which determines not only values of GLA (m^2) but also the component of GLA (CGLA) used by the current tenants or owner-occupiers, as mentioned briefly at the start of this overview of the method. Untenanted vacancy (GLAU) can be calculated by comparing the Component of GLA (CGLA) against the GLA data for each suite in a building. If the CGLA (space used) is equal to GLA, the area is considered fully occupied and has no greyspace vacancy. Likewise, if CGLA values are equal to 0 m^2, the space is deemed untenanted and vacant. Greyspace (VG), therefore, can be derived from understanding that when CGLA values are greater than 0 m^2 but less than its GLA value, space is partially occupied (tenanted) but also underused. The formulae for calculating Untenanted and greyspace vacancy on a building-by-building basis are shown by formula 4 in Figure 4.3.

An occupancy rate (OR %) must be calculated for each building before vacancy rates (VR %) can be determined. A vacancy rate (VR) is then calculated for each building by finding the OR (%) inverse. For this, see formula 4 in Figure 4.3. It is also helpful to understand vacancy in terms of area (m²) by calculating Vacant Area (VA).

4.3.2 Myth-busting findings from VVAM

VVAM findings have discovered that a nuanced understanding of vacancy is valuable. Its conclusions can challenge in-grained perceptions about vacancy rates and reactive policy solutions to resolve perceived vacancy. To illustrate the value of a nuanced understanding of vacancy, beyond a reliance on simplistic aggregated vacancy rates, this section of the chapter will discuss an overview of what was found for a city that has suffered from high average vacancy rates in office buildings for many decades. The city is Adelaide, an Australian state capital city with a declining population (ABS, 2018). The full findings are detailed in Armstrong (2020).

For Adelaide, although the average vacancy rates were considered high across the population, VVAM found that across the largest possible sample of office buildings (n = 118), the CBD had no wholly empty office buildings at the peak of the vacancy rates in 2017. The size of a building was found to be a factor in the presence of high levels of vacant space, disclosing that buildings with larger office-use floor areas tended to have higher vacancy (oVR) (Armstrong, 2020). The untenanted space was overwhelmingly distributed in pockets of single floor plates or partial floor plates. Of the sample of 118 office buildings, only two large-scale buildings were found to have a vacancy rate of above 50% (oVR) which contained vacancy in stacked floor plates. Most buildings contained pockets of vacancy occupancy scattered in and among occupied floor plates.

Further to this, vacancy was distributed across both the high-quality grades and secondary grade office buildings. The cross-sectional data analysis also detected no discernible 'lump' of vacant space in the secondary grade office buildings. This finding challenges the need to focus only on vacancies in secondary grade building stocks. Further to the 'indigestible lump' concept, the 'sinking stack theory' predicts that vacancy would be concentrated in the secondary grade office buildings (Langston et al., 2008; Ness & Atkinson, 2001; Atkinson, 1988). Data did not bear out this prediction, although a longitudinal study may detect the operation of the sinking stack process.

4.3.3 Resilience capacity strategies

The findings for Adelaide collectively indicated that whole building adaptive reuse (WBAR) is unlikely to be a suitable option to address office building vacancy in Adelaide's city centre. Out of a sample of 118, only ten office buildings had the capacity for mixed-use multi-level adaptive reuse (MUMLAR). To increase the likelihood of MUMLAR as a vacancy solution, building owners and asset managers would first need to consolidate the pockets of vacancy and relocate the occupying business to group the pockets of vacant space together. Mixed-use multiple level adaptive reuse (MUMLAR) is therefore also unlikely as a solution in the short term. VVAM's analysis for the case of Adelaide revealed that partial adaptive reuse (PAR) is the most suitable category of adaptive reuse for the office building vacancy problem in Adelaide in the immediate to short term.

Enabling behaviours to manage and reduce the risk of premature obsolescence by vacancy presence could be picked up by VVAM analysis. The occupancy distribution suggested a lack of space-use consolidation occurring in the large-scale ($GLA_{BUILDING} \geq 3000$ m^2) secondary grade office buildings with a high vacancy rate (oVR $\geq$ 50%). Questions need asking about how active current owners are in adopting behaviours to reduce the risk of their buildings becoming stranded (Muldoon-Smith & Greenhalgh, 2019). Is consolidation a first step in mitigating obsolescence?

4.4 Discussion

The findings from VVAM for the case of Adelaide raise serious questions for other cities about how simplistic understanding of vacancy shapes policy development, particularly when vacancy rates are used to call for regulation reform, such as for planning systems or building regulation's performance standards. At present, urban vacancy is typically addressed through ad hoc public policies reactive to where and when vacancy emerges (Buitelaar et al., 2021). In the case of Adelaide, building regulation, particularly performance standards in Australia's National Construction Code, was problematised by industry stakeholder groups as a critical factor inhibiting greater adaptive reuse uptake (Armstrong, 2020). However, without a nuanced understanding of the types and distribution of vacancy, pressure to reform NCC performance standards is challenging to rebuke by local and state government policymakers. The development of VVAM seeks to enable commitments towards a more anticipative policy framework for sustainable management of buildings, as recommended by Buitelaar et al. (2021).

4.4.1 Beyond untenanted vacancy – a call for a more nuanced understanding of vacancy

Untenanted vacancy is perhaps the vacancy type most familiar and used in calculating vacancy averages by property market analysts. It is also suitable for quantifying vacancies in cross-sectional studies. A key advantage of vacancy types suitable for cross-sectional studies is that they can inform analysis that needs to be 'generally quick, easy, and cheap to perform' (Sedgwick, 2014:2). This efficiency is helpful for policy development with limited resources. Cross-sectional studies are also suitable for providing an understanding of vacancy in a particular 'snapshot' in time, such as when vacancy rates are perceived as being problematically high or low or when cities experience unforeseen acute shock events and increased stress. It should be noted here that there are many types of vacancy valuable to identify when developing solutions to the problem of high vacancy, particularly when developing policy to build capacity for urban resilience.

As highlighted by VVAM, greyspace vacancy is another vacancy type suitable for cross-sectional studies. It is a type of vacancy that can have a significant effect on urban vibrancy and economics. For the case of Adelaide, greyspace surprisingly outweighed untenanted vacancy across both primary and secondary grade office buildings (Armstrong, 2020). This finding highlights both the problem of vacant space for Adelaide and the usefulness of vacancy to evaluate solutions to reduce or mitigate vacancy impacts.

While this chapter acknowledges the importance of looking at different types of vacancy, the reader should also note that an understanding of different vacancy types is an undeveloped field of research at present (Muldoon-Smith & Greenhalgh, 2017). In addition to this, quantifying different vacancy types is messy because using secondary data is limiting and the datasets available dictate the kinds of vacancy that can be quantified.

4.4.2 Adaption is happening but not as we know it

The novelty of adaptive reuse of heritage buildings dominates the field of adaptive reuse research. While these studies are of interest, dominant methods and heritage focus limit opportunities to draw generalisable conclusions about adaptive reuse for the vast majority of buildings in our cities. Methods in adaptive reuse research are also predominantly reliant on small-scale descriptive case studies. There are, however, a small number of studies with quantitative or mixed-method methodologies. These studies examine adaption at a city scale across stocks of existing buildings in Australia (Armstrong, 2020; Wilkinson, 2011), the UK (Muldoon-Smith, 2016), Europe (Huuhka & Kolkwitz, 2021; Remøy, 2010) and Asia (Sing et al., 2019). These studies take a different approach to understanding adaption and have larger sample sizes, contributing potentially generalisable findings.

VVAM highlighted cases of adaptive reuse through understanding the space uses of existing buildings. In Adelaide's case, commercial office buildings had already undergone some level of adaptive reuse (Armstrong, 2020). However, VVAM also highlighted that whilst adaptive reuse is happening in Adelaide, and it was often not whole building adaptive reuse. The adaptive reuse that was happening in Adelaide was of smaller scale pockets – pocket adaptive reuse (PAR) and mixed-use multiple level adaptive reuse (MUMLAR). Where whole building adaptive reuse has occurred, the result was a quite low-quality budget accommodation and did little to improve the quality of the adapted building. This low-key, low-quality, smaller-scale adaptive reuse was alive and well in Adelaide, but not perhaps the kind of adaptive reuse expected or noticed.

4.4.3 Missing vacancy problem and the problem with vacancy framing

While adaption of vacant buildings is a recognised sustainable strategy, without an explicit exploration of vacancy in buildings, adaptive reuse of existing buildings can run the risk of negative consequences. If the vacancy is not fully understood, adaptive reuse and regeneration developments can displace important communities

(Abramson, 2016) and transition of land to new uses/zones due to pressure from more economically attractive property markets (Clifford et al., 2018; Grodach et al., 2017). This highlights that missing vacancy data can result in unintended negative consequences of community displacement or low-quality adaptive reuse.

The analysis of vacancy for Adelaide highlights the need for a deeper and more nuanced understanding of vacancy in existing building populations to assist policy decisions in addressing obsolescence in the built environment. This need aligns with Muldoon-Smith's (2016) study of office building vacancy in the UK. However, despite the power of data to inform policy mechanisms, the problematisation of vacancy and its possible solutions are often not informed by data or research.

Adelaide problem-solution framing can be better understood by applying Bacchi's (2012) approach titled 'What is the Problem Represented (WPR) to be'. Adelaide was considered to have a high level of vacancy in its office buildings. The presence and persistence of this vacancy were presented as evidence of regulatory barriers to adaptive reuse, and adaptive reuse was framed as the solution if regulation could be reduced for existing building adaption. The framing of Adelaide's problem-solution by property industry groups is one example where simplistic vacancy average rates were relied upon, and local and state governments undertook reactive policy work. Bacchi's (2012) approach is useful to unpack how the problem of high vacancy evolved in public debate and the proposed solution to increase adaptive reuse uptake by relaxing the National Construction Code. VVAM offered insights, through analysing fine-grain vacancy data, and showed the distribution of vacancy for Adelaide's office buildings did not lend itself to whole building adaptive reuse.

VVAM is also valuable when triangulating qualitative data on the barriers and enablers of adaptive reuse or other vacancy mitigation strategies offered as potential solutions. For research into adaptive reuse, triangulation is vital because it limits potential bias stemming from qualitative data such as interviews with stakeholders. A range of state and local policy levers can offer financial gain for stakeholders, particularly building owners and developers. These policy levers include regulation dispensations, tax concessions, planning approval exemptions and grants or loans to upgrade existing buildings. Triangulation of qualitative data can limit the bias from stakeholders who seek to financially benefit from encouraging policy action around reducing the cost of building regulation and improving the financial viability of undertaking adaptive reuse. This is important as incentives can have unintended negative consequences, especially if there are calls to reduce compliance with fire safety, disabled access and energy efficiency codes.

4.5 Conclusion

Throughout this chapter, vacancy is positioned as an important indicator of the risk of premature obsolescence in existing buildings. Premature obsolescence leads to an increased risk of demolition, which in turn generates increased consumption of materials and resources sent to landfills. However, the process of obsolescence is poorly studied and documented, with many discussions focusing only on buildings that have reached their terminal condition of long-term structural vacancy and low or zero levels of occupancy. This chapter makes a case for a more nuanced understanding of the vacancy through robust datasets to inform and shape effective policy to mitigate the risk of obsolescence at a much earlier stage of a building's life cycle. This chapter also focused on looking at vacancy at the scale of the whole city, across building stocks, to develop generalisable findings to build resilient cities rather than on a case-by-case basis. It set out a new method, referred to as VVAM, as a practical tool for quantifying vacancy by repurposing data already collected. The reuse of secondary data makes the research process as efficient as possible. The chapter then focused on some myth-busting findings from applying VVAM to a case city located in Australia, Adelaide. The results make the case that vacancy understanding can offer essential insights and challenge prevailing industry views about vacant space and solutions to mitigate vacancy.

The overview of the current understanding of vacant space highlights the significant gaps remaining. These gaps include robust testing of the dominant theory of vacancy and the process of obsolescence known as the 'sinking stack' theory. It highlights that we have not yet developed a consensus of different vacancy types useful as indicators of the risk of premature obsolescence. Currently, we focus predominantly on simplistic aggregated untenanted vacancy rates. However, VVAM offers a step-by-step process to provide fine-grain untenanted vacancy data for every building across a city. VVAM is also the first known study to quantify grey-space vacancy, a hitherto important but hidden vacancy type. Finally, this chapter set out the need to be more critical of the problematisation of vacancy by industry groups, particularly when fine-grain vacancy is not available.

The availability of data to quantify vacancy is a potential limitation of VVAM and should be acknowledged. It is possible to repurpose taxation data if local councils link the rateable value for commercial buildings to their occupancy levels or floor use. Where jurisdictions do not connect commercial building rates with occupancy, other data sources are needed. The research in this book chapter recommends the development of local councils to collect and publish vacancy data for all building populations important to its economic activity.

So, where next for vacancy understanding? To develop an evidence-informed resilience policy, reducing the risk and occurrence of premature obsolescence in cities, we need access to fine-grain vacancy data. We need transparency over how these data are collected and the sample used. We need an analysis of different vacancy types and where they are distributed across other building stocks. We need to start building cross-sectional datasets now, and over time, build longitudinal fine-grain understanding. This exploration needs to be done not only for cities with high vacancy levels but also for cities with unhealthy levels of low vacancy, as low vacancy puts pressure on important land uses that are less profitable and play a vital role in resilient cities. Policy reliance on simplistic, aggregated vacancy rates will tend to be reactive, and for resilience, we need to move to anticipative policies. The need for resilience and anticipative policies is highlighted by the vacancy scenarios provided at the start of this chapter. A resilient city can weather the stresses and acute shocks with greater levels of social equity, shorter periods of economic decline, and reducing premature obsolescence in cities.

References

Abdullah, M. S. M., Suratkon, A. and Mohamad, S. B. H. S., 2020. Criteria for adaptive reuse of heritage shop houses towards sustainable urban development. *International Journal of Sustainable Construction Engineering and Technology*, *11*(1), pp. 42–52.

Abramson, D. M., 2016. *Obsolescence: An architectural history*. University of Chicago Press, Chicago, IL.

Aigwi, I. E., Phipps, R., Ingham, J. and Filippova, O., 2020. Characterisation of adaptive reuse stakeholders and the effectiveness of collaborative rationality towards building resilient urban areas. *Systemic Practice and Action Research*, *33*(3), pp. 1–11. https://doi.org/10.1007/s11213-020-09521-0.

Armstrong, G., 2020. The adaptive reuse predicament: An investigation into whether building regulation is a key barrier to adaptive reuse of vacant office buildings (Doctoral dissertation). Available from http://hdl.handle.net/2440/129492.

Armstrong, G., Soebarto, V. and Zuo, J., 2021. Vacancy visual analytics method: Evaluating adaptive reuse as an urban regeneration strategy through understanding vacancy. *Cities*, *115*, p. 103220.

Atkinson, B., 1988. Urban ideals and the mechanism of renewal. RAIA Conference, Sydney. RAIA.

Australian Bureau of Statistics (ABS), 2018. Summary for South Australia. In *3222.0 -Population Projections, Australia, 2017–2066*. Available from: www.abs.gov.au/ausstats/abs@.nsf/7d12b0f6763c78caca257061001cc588/fd938fe1fe00d597ca257c2e001724b5!Open Document. Accessed 22 November 2019.

Australian Centre of Excellence for Local Government (ACELG), 2013. Knowledge city. Available from https://apo.org.au/node/33925.

Australian Property Institute (API), 2017a. *Technical Paper Information – Methods of Measurement*. Available from www.api.org.au. Accessed 26 March 2019.

Australian Property Institute (API), April 2017b. Investor insight: Office market outlook. Available from www.investa.com.au/WWW_Investa/media/Resources/April-2017-Office-Market-Report.pdf?ext=.pdf&alt=&fpl=&fpt=&fpr=&fpb=&guid=0fd9d529-0db5-471f-9dbc-a3ee38ad8bf0&lgW=0&lgH=0&mdW=680&mdH=0&smW=480 &smH=0. Accessed 3 October 2017.

Bacchi, C., 2012. Introducing the 'what's the problem represented to be?' Approach. In A. Bletsas and C. Beasley (Eds.), *Engaging with Carol Bacchi: Strategic Interventions and Exchanges* (pp. 21–24). University of Adelaide Press, Adelaide.

Barras, R., 2009. *Building Cycles: Growth and Instability* (Vol. 27). John Wiley & Sons, New York.

Barras, R. and Clark, P., 1996. Obsolescence and performance in the Central London office market. *Journal of Property Valuation and Investment*, *14*(4), pp. 63–78.

Best, R., Langston, C. A. and De Valence, G. (Eds.), 2003. *Workplace Strategies and Facilities Management*. Routledge, London.

Bodfish, H. M., 1931. Practical uses of vacancy statistics. *Journal of the American Statistical Association*, *26*(173A), pp. 53–57.

Bottero, M., D'Alpaos, C. and Oppio, A., 2019. Ranking of adaptive reuse strategies for abandoned industrial heritage in vulnerable contexts: A multiple criteria decision aiding approach. *Sustainability*, *11*(3), p. 785. https://doi.org/10.3390/su11030785.

Buitelaar, E., Moroni, S. and De Franco, A., 2021. Building obsolescence in the evolving city. Reframing property vacancy and abandonment in the light of urban dynamics and complexity. *Cities*, *108*(2021), p. 102964. https://doi.org/10.1016/j.cities.2020.102964.

Burkholder, S., 2012. The new ecology of vacancy: Rethinking land use in shrinking cities. *Sustainability*, 4(6), pp. 1154–1172. https://doi.org/10.3390/su4061154.

City of Melbourne (CoM), 2013. Knowledge city strategy 2014–2018. Available from www.melbourne.vic.gov.au/about-council/vision-goals/knowledge-city/Pages/knowledge-city-strategy.aspx. Accessed 13 February 2022.

City of Sydney, 2017. Floor space and employment survey. Available from www.cityofsydney.nsw.gov.au/surveys-case-studies-reports/floor-space-employment-survey-2017. Accessed 13 February 2022.

Clifford, B., Ferm, J., Livingstone, N. and Canelas, P., 2018. *Assessing the Impacts of Extending Permitted Development Rights to Office-to-Residential Change of Use in England.* Royal Institution of Chartered Surveyors, London.

Felstead, A. and Reuschke, D., 2020. *Homeworking in the UK: Before and During the 2020 Lockdown*, WISERD Report. Wales Institute of Social and Economic Research, Cardiff. Available from https://wiscrd.ac.uk/publications/homeworking-uk-and-during-2020-lockdown. Accessed 13 February 2022.

Fenton, M., 2020. 2020 wrap up for NSW and predictions for 2021. *Savills Blog*. Available from www.savills.com.au/blog/article/198762/australia-articles/nsw-2020-wrap-up-and-predictions-for-2021.aspx. Accessed 13 February 2022.

Fitzgerald, K., 2020. *Speculative Vacancies 10: A Persistent Puzzle, the Study of Melbourne's Vacant Land and Housing.* Prosper Australia. Available from www.prosper.org.au/2020/12/the-persistent-puzzle-speculative-vacancies-10/. Accessed 13 February 2022.

Foster, G., January 2020. Circular economy strategies for adaptive reuse of cultural heritage buildings to reduce environmental impacts. Resources, *Conservation and Recycling, 152*, p. 104507. https://doi.org/10.1016/j.resconrec.2019.104507.

Goldsmith, S. and Crawford, S., 2014. *The responsive city: Engaging Communities Through Data-Smart Governance.* John Wiley & Sons, New York.

Greenhalgh, P. and Muldoon-Smith, K., 2017. *Mitigating Office Obsolescence: The Agile Future.* British Council for Offices, London.

Grodach, C., O'Connor, J. and Gibson, C., 2017. Manufacturing and cultural production: Towards a progressive policy agenda for the cultural economy. *City, Culture and Society, 10*, pp. 17–25. http://dx.doi.org/10.1016/j.ccs.2017.04.003.

Hammond, E., 2013. Surplus commercial property leases worth up to £75bn. *Financial Times.* Available from www.ft.com/content/d25b515e-f38b-11e2-b25a-00144feabdc0. Accessed 4 November 2019.

Holden, G., 2018. Top-up: Urban resilience through additions to the tops of city buildings. In S. J. Wilkinson and H. Remøy (Eds.), *Building Urban Resilience Through Change of Use* (pp. 105–130). John Wiley & Sons, London.

Huuhka, S. and Kolkwitz, M., 2021. Stocks and flows of buildings: Analysis of existing, demolished, and constructed buildings in Tampere, Finland, 2000–2018. *Journal of Industrial Ecology, 2021*(25), pp. 948–960. https://doi.org/10.1111/jiec.13107.

Johnston, M. P., 2017. Secondary data analysis: A method of which the time has come. *Qualitative and Quantitative Methods in Libraries, 3*(3), pp. 619–626.

Langston, C., Wong, F. K., Hui, E. C. and Shen, L. Y., 2008. Strategic assessment of building adaptive reuse opportunities in Hong Kong. *Building and Environment, 43*(10), pp. 1709–1718. https://doi.org/10.1016/j.buildenv.2007.10.017.

Lavis, J. N., Posada, F. B., Haines, A. and Osei, E., 2004. Use of research to inform public policymaking. *The Lancet, 364*(9445), pp. 1615–1621. https://doi.org/10.1016/s0140-6736(04)17317-0.

Lesneski, T., 2011. Big box libraries: Beyond restocking the shelves with books. *New Library World, 112*(9/10), pp. 395–405. https://doi.org/10.1108/03074801111181996.

Lynch, N., 2021. Remaking the obsolete: Critical geographies of contemporary adaptive reuse. *Geography Compass*, 16(1), e12605. https://doi.org/10.1111/gec3.12605.

Muldoon-Smith, K., 2016. *Taking stock: An investigation into the nature, scale and location of secondary commercial office vacancy in the UK and an appraisal of the various strategies and opportunities for its management and amelioration* (Doctoral thesis), Northumbria University.

Muldoon-Smith, K. and Greenhalgh, P., 2017. Situations vacant: A conceptual framework for commercial real estate vacancy [paper presented]. 24th Annual European Real Estate Society Conference. Conference. Delft, Netherlands, ERES. http://dx.doi.org/10.15396/eres2017_341.

Muldoon-Smith, K. and Greenhalgh, P., 2019. Suspect foundations: Developing an understanding of climate-related stranded assets in the global real estate sector. *Energy Research & Social Science*, 54, pp. 60–67. https://doi.org/10.1016/j.erss.2019.03.013.

Ness, D. A., 2002. Accounting for the effects of oversupply of commercial property (Doctoral thesis), University of South Australia.

Ness, D. A. and Atkinson, B., 2001. *Design Guide 11: Re-use/Upgrading of Existing Building Stock*. Australian Council of Building Design Professions Ltd, Melbourne.

Petit, C., Wentz, E., Randolph, B., Sanderson, D., Kelly, F., Beevers, S. and Reades, J., 2020. Tackling the challenge of growing cities: An informed urbanisation approach. In S. Hawken, H. Han, and C. Pettit (Eds.), *Open Cities | Open Data* (pp. 197–219). Palgrave Macmillan, Singapore. https://doi.org/10.1007/978-981-13-6605-5_9.

Pottie-Sherman, Y., 2020. Rust and reinvention: Im/migration and urban change in the American Rust Belt. *Geography Compass*, 14(3), p. e12482. https://doi.org/10.1111/gec3.12482.

Remøy, H. and Street, E., 2018. 'The dynamics of "post-crisis" spatial planning: A comparative study of office conversion policies in England and The Netherlands. *Land Use Policy*, 77(2018), pp. 811–820. https://doi.org/10.1016/j.landusepol.2016.12.005.

Remøy, H. and van der Voordt, T., 2014. Adaptive reuse of office buildings into housing: Opportunities and risks. *Building Research & Information*, 42(3), pp. 381–390. https://doi.org/10.1080/09613218.2014.865922.

Remøy, H. T., 2010. *Out of Office: A Study on the Cause of Office Vacancy and Transformation as a Means to Cope and Prevent*. IOS Press, Amsterdam.

Roberts, E. and Carter, H. C., 2020. Making the Case for centralized dementia care through adaptive reuse in the time of COVID-19. *Inquiry: A Journal of Medical Care Organization, Provision and Financing*, 57, pp. 1–6. https://doi.org/10.1177%2F0046958020969305.

Rosenthal, C. L. and Rothenberg, L. E., 2021. The exodus from New York due to COVID-19. *CPA Journal*. Available from www.cpajournal.com/2021/09/22/icymi-the-exodus-from-new-york-due-to-covid-19/. Accessed 13 February 2022.

Saniga, A., 2012. *Making Landscape Architecture in Australia*. UNSW Press, Sydney.

Savills, 2020. Impact of Covid-19 on European retail. *Spotlight Savills Research*. Available from https://pdf.euro.savills.co.uk/portugal/spotlight-european-retail-sep-20.pdf. Accessed 13 February 2022.

Sedgwick, P., 2014. Cross sectional studies: Advantages and disadvantages. *BMJ*, 348, pp. 2276–2277. https://doi.org/10.1136/bmj.g2276.

Sing, M. C., Love, P. E. and Liu, H. J., 2019. Rehabilitation of existing building stock: A system dynamics model to support policy development. *Cities*, 87, pp. 142–152. https://doi.org/10.1016/j.cities.2018.09.018.

Turok, I., Seeliger, L. and Visagie, J., 2019. Restoring the core? Central city decline and transformation in the South. *Progress in Planning*, 144, p. 100434. https://doi.org/10.1016/j.progress.2019.100434.

Wilkinson, S. J., 2011. The relationship between building adaptation and property attributes (Doctoral thesis), Deakin University, Australia.

Wilkinson, S. J., 2018. Sustainable office retrofit in Melbourne. In *Routledge Handbook of Sustainable Real Estate* (pp. 70–82). Routledge, London.

Wilkinson, S. J. and Remøy, H. (Eds.), 2018. *Building Urban Resilience Through Change of Use*. John Wiley & Sons, London.

Winkler, T., 2013. Why won't downtown Johannesburg 'Regenerate'? Reassessing Hillbrow as a case example. *Urban Forum, 24*, pp. 309–324. https://doi.org/10.1007/s12132-012-9178-5.

Wolff, M. and Wiechmann, T., 2014. Indicators to measure shrinking cities. In Cristina Martinez-Fernandez, Sylvie Fol, Tamara Weyman and Sako Musterd (Eds.), *A Conceptual Framework for Shrinking Cities* (pp. 1–19). COST Action TU0803: Cities Re-growing Smaller (CIRES).

Part 2

The what

Exploring solutions

5 A governance response

From coercion to persuasion to embracing diversity?

Jeroen van der Heijden

5.1 Introduction

Seeking to achieve building retrofits, government responses can be, and have been, coercive, persuasive, or both, and have ranged from punitive tax regimes and statutory requirements to "nudge" techniques and voluntary programmes. This chapter analyses a range of measures in different jurisdictions and across the spectrum of interventions (such as taxes, certification requirements, statutory obligations, and economic incentives). It assesses whether such measures are sufficient in light of the great urgency posed by global challenges, including climate change (adaptation and mitigation) and the Covid-19 pandemic (health and well-being). It argues that not enough is being done to shift the pendulum from coercive to persuasive techniques, and suggests ways in which governments should seek higher levels of effectiveness through an overhaul of the building regulatory system. This overhaul would involve combining coercive and persuasive interventions and targeting different groups of property owners and users with tailored regulatory and governance interventions.

A well-known approach to governing the built environment is to introduce coercive building regulation and building codes. While such direct regulatory interventions have proved to be relatively successful in the past, the introduction of new statutory regulation has significant drawbacks for achieving building retrofits, urban climate adaptation, and improved health and well-being of the built environment more broadly (a topic that has received attention due to the Covid-19 pandemic).

Three challenges to coercive regulation stand out:

- It is a slow working tool. It often takes a long time to develop and implement legislation and regulation and even longer for it to have an effect. Ever-accelerating climate change and its consequences may outpace regulatory solutions (Romero-Lankao et al., 2018), and "shock-like" events such as the Covid-19 pandemic typically do outpace them. Moreover, in rapidly developing economies that are experiencing construction booms, the slow development of regulation may result in building stock that is not future proof.

DOI: 10.1201/9781003023975-7

- It is ineffectual for achieving building retrofits. Existing buildings are typically exempted from new regulatory requirements, a process known as "grandfathering". Given that most cities in developed economies transform at less than 2% per year, it may take 40–70 years for new regulations to transform all buildings and infrastructure (van der Heijden, 2015b).
- It requires high levels of institutional capital if it is to be effective. Typically, it involves a staged process of building proposal assessments and permits, on-site construction inspections, completion and occupation checks, and so on. Developing economies may lack the institutional capital to carry out these tasks effectively, and experience with building regulation and control in developed economies indicates high levels of non-compliance with building regulation even when such tasks are carried out (van der Heijden, 2009).

To overcome these challenges, governments around the world have begun moving from traditional, top-down, coercive interventions to more persuasive interventions that are often developed in collaboration with stakeholders. This shift from coercive to persuasive interventions has been documented for well over three decades (van der Heijden, 2014). While high hopes were expressed initially about the ability of persuasive interventions to accelerate the transition to more climate resilient, healthy, and prosperous buildings and cities (Barber, 2013; Sassen, 2015), scholars have more recently begun to highlight the systematic shortfalls (Sanchez et al., 2018; Solecki et al., 2018; van der Heijden et al., 2019). In response, scholars have begun to call for hybrid solutions: mixing and matching of coercive and persuasive interventions.

This chapter maps, explores, and interrogates the shift from coercive regulation to persuasive intervention by governments with the main focus on developed economies. After discussing some examples of persuasive governance interventions for building retrofits (mainly from the area of urban climate governance), and their opportunities and limits, the call for hybrid solutions is extended by requesting a broader rethinking of the regulatory and governance system for building retrofits.

5.2 From coercive to persuasive governance interventions

The origins of building regulation and its enforcement can be traced back thousands of years and have been documented extensively elsewhere (Leon & Rossberg, 2012; Meacham, 2016; van der Heijden, 2014). For the purposes of this chapter, it is of relevance to understand the major shifts in the regulation of buildings – and regulatory governance more broadly – that have occurred around the globe since the 1980s.

5.2.1 General governance trends

Stepping back from the topic of building retrofits for a moment, it is important to acknowledge that since the 1980s, substantial changes have occurred in public governance and regulation in a wide variety of policy areas. Scholars often describe this as a shift "from government to governance", indicating that in the governing of society the government no longer takes central stage (Rhodes, 1996, 2007). This shift includes the privatisation of public service delivery in the 1980s and 1990s in the US, the UK, and elsewhere, and the related shifts in deregulation and reregulation (Hodge, 2000). It includes the embracing of new public management practices by governments in that same period (Osborne & Gaebler, 1992), as well as the rise in the power and size of companies which now often eclipse the power and size of local and national governments (Carroll, 2007; Cowling & Tomlinson, 2005). It includes the ongoing collaboration between government, the private sector, and civil society in the development and implementation of public policy and regulation that has grown exponentially since the early 2000s (Chhotray & Stoker, 2010). Finally, it includes the growth of self-governing and polycentric initiatives by firms, citizens, and municipalities, in areas such as climate change responses, healthy lifestyles, and the future of democracy, which have become dominant since the 2010s (Jordan & Huitema, 2019).

5.2.2 Ways in which persuasive governance is displayed

In the regulation of buildings, the shift "from government to governance" manifests itself most clearly in the type of building regulations and codes that we see today, their enforcement, and a variety of other-than-regulatory interventions that governments have put in place to achieve safe, healthy, and sustainable buildings.

5.2.2.1 From prescription to performance

One of these manifestations is a move away from traditional prescriptive building regulation to performance and goal-oriented regulation. Prescriptive building regulation seeks to prevent harmful events, for instance the collapse of a building, by stating the exact requirements that must be met by the parts of the building, its construction, and its design process. Prescriptive standards are typically criticised for lacking flexibility, hampering innovation, and requiring a level of technical expertise that governments often lack (Meacham et al., 2005). Performance and goal-oriented building regulation partly overcomes this criticism. It specifies the performance or goal of a good or service but not how that performance or goal is to be achieved. Such regulation is normally considered to give those regulated – whether they are property developers, building owners, or occupants – an incentive to find a solution that is both effective in terms of meeting the standard and efficient in terms of costs (May, 2011). However, in the wake of major disasters such as the 2017 Grenfell Tower fire in London, UK, some have begun to rethink performance and goal-oriented regulation (Law & Butterworth, 2019).

5.2.2.2 From governments to NGOs

Another way in which the shift is manifested is in a move away from a traditional model in which a (national) government sets building regulations and a (local) government enforces them. Around the world, we witness the ongoing embracing by governments of building regulations and standards developed by non-governmental organisations. A typical example is the reference to voluntary energy and other sustainable building standards developed by LEED (Leadership in Energy and Environmental Design, discussed further in what follows) by a variety of municipalities in the US. In some municipalities, compliance with LEED standards is deemed to be equivalent to compliance with mandatory local government building regulations; in other municipalities, local governments have directly made LEED standards part of their local building regulatory regime (van der Heijden, 2015a). In a similar vein, governments around the world have opened up their monopolies of building code enforcement, and now allow private firms and individuals to carry out building plan checks and on-site construction inspections, and sometimes even to issue construction and occupancy permits. In short, in some countries governments have become facilitators or orchestrators of building regulatory and governance systems, rather than acting as the "trinity" of legislator, judge, and executor of these systems (van der Heijden, 2017a).

5.2.2.3 From negative to positive incentives

The third manifestation is the embracing of a broad set of governance interventions – other than mandatory building regulations – to achieve safe, healthy, and sustainable buildings. For a long time, a central assumption underpinning government regulation was that people comply with regulations because they fear the consequences of being found to have violated them. Think of fines and prison terms imposed on violators of building codes, and sometimes the demolition of buildings found to violate building regulations. Since the 1980s, insights from the behavioural sciences have been challenging this assumption by showing that people respond to a wide variety of positive and negative incentives and have a wide variety of motivations to comply with regulation (Drahos, 2017). Inspired by such insights, governments have begun developing targeted incentive schemes, such as tax breaks for energy-efficient buildings, low-interest loans for innovative construction techniques, and carbon-trading schemes for office buildings (Romero-Lankao et al., 2018).

5.2.2.4 Nudge techniques and voluntary programmes

Other examples include nudge-type interventions that seek to address people's heuristics and biases. For example, if households are informed about how their energy consumption compares with that of their neighbours, they are likely to reduce their energy consumption voluntarily if it is higher than that of their neighbours (van der Heijden, 2020b). In the slipstream of these government interventions, there is a growth of non-government initiatives and voluntary programmes that seek to incentivise developers, property owners, and occupants to move beyond mere bottom-line compliance with mandatory government building regulations (van der Heijden, 2017c) by, for example, introducing certification for buildings with beyond-compliance accessibility for people with disabilities, or reducing insurance premiums for buildings with beyond-compliance earthquake resilience (Egbelakin et al., 2017). Often such "novel" governance interventions are developed in experimental settings in which governments work with stakeholders to test and fine-tune interventions before they are rolled out on a large scale (van der Heijden, 2015c; van der Heijden & Hong, 2020).

5.2.2.5 Summary

The shift from government to governance in building regulation implies, in practice, that governments have moved away from mandatory regulation as a "one size fits all" solution that will produce safe, healthy, and sustainable buildings by force. Rather, they (and others) have embraced a variety of persuasive governance interventions to achieve these goals. But are these persuasive interventions effective for achieving building retrofits and, more broadly, urban climate mitigation?

5.3 Persuasion for building retrofits: examples and opportunities

What follows is a brief overview of some of the dominant persuasive governance interventions which have become popular since the early 2000s. This overview is by no means exhaustive – more extensive overviews are available elsewhere (e.g. Bulkeley, 2010; Hoffmann, 2011; Romero-Lankao et al., 2018; van der Heijden, 2014, 2017c). This overview is meant to give the reader a snapshot of the variety of interventions in place and to illustrate how persuasive governance interventions have been implemented to achieve building retrofits and more generally to drive the transition towards resilient built environments. There are a number of ways this has happened.

5.3.1 Alternative forms of financing

Around the globe, property developers, owners, and users are incentivised to retrofit their buildings by helping them to obtain finance (van der Heijden, 2017b). For example, in 2010, the government of San Francisco introduced the Voluntary Retrofit Ordinance (VRO). The ordinance sought to accelerate the earthquake strengthening of soft storey buildings – buildings that have one level that is considerably more flexible than the one below or above it. About 60,000 residents of San Francisco live in soft storey buildings (Baldridge, 2012). The VRO provides for expedited building permits, waives permit fees and plan review fees, and provides exemptions from future mandatory seismic upgrades for a period of 15 years. However, these incentives are considered too weak to be effective (Baldridge, 2012); the fees waived are often less than 3% of the total retrofit cost. In 2010, for example, only 26 buildings participated in the VRO (Baldridge, 2012), and in 2013, the government of San Francisco passed a mandatory retrofit ordinance for these buildings (Chakos & Zoback, 2018). To support property owners with soft storey retrofit improvements, the government of San Francisco is providing public financing through PACE programmes.[1] PACE, or Property Assessed Clean Energy, programmes are available throughout the US.[2] Initially, PACE financing was made available for building energy efficiency improvements for existing buildings, but it now also includes financial support for risk mitigation and water conservation improvements. PACE is not a loan or a subsidy but a tailored financing agreement between a local government and a property owner. Once it has entered into an agreement, the local government issues a bond on behalf of the property owner. The bond can be purchased by a third-party finance provider. After obtaining funds, the government of San Francisco supplies these to the property owner, who uses the money for retrofits and upgrades in accordance with the agreement. The government of San Francisco recoups these funds – with interest – through an additional levy (a "non-ad valorem assessment") on the property, and it pays back the funds and interest to the third-party finance provider. This means that the duty to repay the funds moves to the new property owner when the ownership of the building changes (van der Heijden, 2018).

5.3.2 Market-based incentives

Governments may choose to incentivise property developers, owners, and users to retrofit their buildings by introducing positive or negative market-based incentives. Two examples serve to illustrate this approach:

- Tokyo's Cap-and-Trade initiative seeks to reduce greenhouse gas emissions from office buildings (Trencher et al., 2016). Under the initiative, a maximum limit for greenhouse gas emissions is set for an individual building or facility (a "cap"). For many existing buildings, the limit is lower than their business-as-usual emissions. To achieve their set limit, property owners can choose to reduce their energy consumption, to retrofit their building, or to purchase renewable energy. However, they can also choose to go beyond their set limit and reduce their property's emissions below the cap. If they do this, they gain a surplus of emission rights that they can sell to other property owners who cannot or will not meet their cap and who prefer to purchase permits from others to offset their emissions.
- The My Safe Florida Home programme (MSFH) is linked to the State of Florida's Windstorm Mitigation Credits (WMCs). It seeks to improve the structural safety of existing homes, particularly so that they are better able to withstand hurricanes. In the early 2000s, the Florida legislature began to require insurance companies to offer WMCs in insurance policies. Homes are awarded WMCs based on their capacity to withstand windstorms – the higher this capacity, the higher the number of credits awarded – and insurers give home insurance discounts for homes that have a higher number of credits. The expectation was that owners of weaker homes, in particular, would carry out (voluntary) retrofits to reap these insurance benefits (Young et al., 2012). The initial challenge of the WMCs was that homeowners had to pay for an inspection to get their credits. Between 2006 and 2009, the Florida government had MSFH in place to address this challenge. Under MSFH, homeowners could have a free inspection of their home and were eligible for home

improvement grants of up to US$5,000 on a dollar-for-dollar matching basis for homes with insured values up to US$300,000. Approximately 400,000 inspections were carried out and 35,000 grants were provided under MSFH. A key lesson learnt from the programme is that homeowners will engage in more mitigation if governments subsidise the costs (Carson et al., 2013). Programmes like these are, however, often too expensive to be maintained by a government in the long run. The MSFH was terminated within three years as a result of budget constraints.

5.3.3 Information provision

Providing prospective buyers and tenants with information about the relative performance of a property can also provide an incentive. Again two examples are provided:

- Since 2009, European member states have had to comply with the European Energy Performance of Buildings Directive (EPDB). This requires the introduction of Energy Performance Certificates (EPCs) that indicate the energy performance of a building, or building unit, calculated according to a methodology stipulated in the EPDB. EPCs must be issued when a building is constructed, sold, or let. Member states are given flexibility to adjust the EPC to their national context (for an example, see Sayce & Hossain, 2020). EPCs must at least indicate the energy performance in a labelled class – for example, "A" indicating low energy consumption and "G" indicating high energy consumption. No minimum requirements are set by the European Commission for the energy performance of existing buildings. It is anticipated that, over time, consumers (buyers and tenants) will choose buildings with better energy performance over those with poorer performance, thus forcing the owners of existing buildings to retrofit buildings with poor energy performance (van der Heijden & Van Bueren, 2013).
- In 2016, the New Zealand government introduced the Building (Earthquake-prone Buildings) Amendment Act. This Act tightened the requirements for the earthquake strengthening of existing buildings.[3] It also introduced a building rating system for earthquake risk – this is a six-point scale, ranging from the minimum grade "E" (very high risk) to the maximum grade "A+" (low risk). Under the Act, all existing buildings are deemed to meet at least a base level in the latest New Zealand national building standard (NBS). This base level is 34% NBS (grade "C"), and new buildings must meet 100% NBS (grade "A"; buildings that outperform the NBS are graded "A+"). It is expected that the rating will increase the call for earthquake-proof buildings. Buildings currently not meeting the base level must be retrofitted or demolished, and their owners may be fined for non-compliance. Indeed, government organisations and major commercial tenants require that the buildings they lease meet at least 67% NBS (grade "B"). This is a strong incentive for commercial property owners to retrofit their buildings beyond the minimum requirements if they wish to have these organisations as their tenants (Filippova & Noy, 2020).

5.3.4 Voluntary benchmarking

Another range of market-based tools that has mushroomed since the 1990s is voluntary benchmarking to persuade property owners to retrofit their existing buildings. They include a variety of voluntary classification and rating tools, such as the international LEED and BREEAM (the BRE Environmental Assessment Method), Green Star in the Asia-Pacific, and local tools such as DGNB (Deutsche Gütesiegel Nachhaltiges Bauen) in Germany, GreenRE (Green Real Estate) in Malaysia, and BEAM plus (Building Environmental Assessment Method) in Hong Kong. Whilst these tools are voluntary, they have often been developed with support from (national) governments (van der Heijden, 2014). Their central idea is simple and elegant: by ranking a building in a certain class, that is, through its performance in terms of energy, water, and material use, as well as its structural safety, it can easily be compared to that of other buildings of the same class – at least in theory (the words "benchmarking", "rating", and "labelling" are often used interchangeably in this context; though they refer to slightly different approaches to classification; see Pérez-Lombard et al., 2009). For existing buildings, two types of assessment are often found: assessment of the (retrofit) design of the building, and assessment of the operational phase of the building (Hyde, 2013).

These voluntary benchmarking tools typically set higher requirements than mandatory government regulations, and property owners are provided with a certificate if they meet the requirements. To be certified, a building (or building plan or construction work) is assessed against a series of predefined regulations. Credits are awarded for each regulatory requirement met, and the greater the number of credits awarded, the higher the classification of the building. For developers, investors, property owners, and tenants it is easy to understand that, on a scale from low performing to high performing, say one to five stars or bronze to gold, a five-star or gold class building is better than a one-star or bronze class building. The expectation is that consumers (property buyers and tenants) will pay a premium for buildings with a high-level certification, which in turn will encourage investors and developers to provide such buildings.

5.4 Persuasion for building retrofits: limits and constraints

Whilst the growth of these persuasive interventions is laudable, they are not a panacea for the governance challenges introduced at the start of this chapter. In what follows, a few of the limitations that are repeatedly mentioned in the literature are discussed.

5.4.1 Challenges across the board

Initiatives that seek to provide alternative forms of financing such as San Francisco's VRO and PACE in the US (see Section 8.3.2) are highly vulnerable to economic swings, which reduce their effectiveness (Baldridge, 2012; Sayce & Hossain, 2020). The VRO had to be brought to a conclusion because of budget constraints, and PACE has also faced considerable financial blows in the past (Sichtermann, 2011).

Performance grants such as MSFH in Florida often indicate that property owners will only carry out retrofits if governments subsidise a large part of the costs. This leads to challenges of (un)desirable redistribution of general taxes to, often, middle-class property owners (van der Heijden, 2017b). The EPCs in Europe have not yet resulted in a substantial increase in the energy efficiency of existing buildings – and in many member states the implementation of the EPCs is flawed, in part because effective enforcement is missing (BPIE, 2015, 2019; Sayce & Hossain, 2020). Likewise, the New Zealand rating system for earthquake-prone buildings may have affected the "the top end of town" because of the requirements set by major tenants, but concerns have been expressed about the other parts of the market. Property owners of (very) high-risk buildings (the lowest grades) may find it difficult to obtain mortgages from banks for their building retrofits, and they may have difficulty in getting insurance for their buildings – which is particularly problematic for homeowners (McRea et al., 2018). Finally, Tokyo's Cap-and-Trade initiative is a local success, but whilst other cities (within and outside Japan) have tried to replicate it, they have not been able to do so (Trencher & van der Heijden, 2019).

Because of the large number of voluntary benchmarking and classification tools available and the differences between them in certification requirements and assessment, it may be difficult for buyers and tenants to assess their value. A building that achieves a top score with one tool does not necessarily also achieve a top score with another tool – thus, property owners have ample opportunity to shop for a tool that meets their own needs rather than the needs of the future owner or tenant of their property (Hyde, 2013; van der Heijden, 2014).

5.4.2 Too little, too late?

These persuasive interventions often build on, or are a derivative of, urban *sustainability* and climate *mitigation* interventions. For example, PACE was initially launched to incentivise property owners to increase the energy efficiency of their buildings, and only later could it be used to support the structural strengthening of existing buildings. In practice, this implies that many of these persuasive interventions focus on increasing building resource efficiency, whereas the *retrofitting* of buildings requires a much larger set of interventions – including seismic strengthening, weather tightening, and age proofing. In addition, voluntary benchmarking tools were mostly introduced in the 1990s for (and were mostly applied to) "new" buildings, and it was only in the 2010s that certification for "in use" and "existing" buildings was introduced – for example, LEED for Operations and Maintenance[4] or BREEAM In-Use.[5] In short, these persuasive interventions are still "too little" (and often "too late") to accelerate building retrofits at high speed and on a large scale (van der Heijden, 2014).

These persuasive interventions generally attract "early movers" but not the majority and the laggards in the market (Rogers et al., 2005). Early movers are those individuals and organisations that recognise the risks that come from not taking action (here: the risks of not retrofitting their buildings) and the opportunities provided by interventions (here: access to funding, reduced time to obtain permits, increased market attraction of their buildings, and so on). They are also often better placed to act compared to the majority and the laggards in the market. For example, they may have some funds available to pay the upfront costs required for a subsidy or a performance-based tax break, or they may simply be a little more skilled in applying for support. Early movers represent, however, only a fraction of social sub-groups. In practice, this implies that many of these persuasive interventions only attract a very small (often single digit) percentage of the market they target (van der Heijden, 2020a). The majority of the market (such as parents, and small and medium-sized firms) often need more persuasion before they engage in building retrofits, and the laggards tend not to move until they are mandated to do so because, for example, they worry that they will not be able to recoup the costs of a retrofit over the time they own or occupy the building (van der Heijden, 2017c).

Thus, to round up this brief discussion of examples, opportunities, limits, and challenges (summed up in Table 5.1), we can conclude that persuasive governance

Table 5.1 Governance interventions for building retrofits: a brief summary

Intervention	Example	Opportunities	Constraints
Coercive regulation.	Building energy efficiency codes.	• Backed by law. • Mandatory for all.	• Slow working. • High enforcement costs. • Inflexible. • Does not inspire movement beyond bottom-line compliance. • Typically exempts existing buildings from new requirements.
Positive incentives.	Tax breaks for low energy buildings.	• Economic pressure to comply. • Inspires movement beyond minimum compliance.	• Slow working. • High enforcement costs. • Fairness of rewards may be contested. • Possible low response due to "irrational" behaviour.
	Government-supported access to finance, e.g. government bonds issued on behalf of property owners.	• Direct link between intervention and outcome. • Inspires movement beyond minimum compliance.	• May require (time intensive) change in legislation. • Fairness of rewards may be contested. • Possible low response due to "irrational" behaviour. • Vulnerable to economic swings.

Intervention	Example	Opportunities	Constraints
	Carbon cap-and-trade.	• Freedom of choice in how to comply. • Rewards those who move beyond minimum compliance. • Adheres to "polluter pays" principle.	• May require (time intensive) change in legislation. • Expensive for governments to maintain in the long run. • Effective enforcement required to ensure that allowances maintain their value. • Redistributional fairness may be contested.
Nudging.	Providing comparative data about household energy consumption.	• Addresses "irrational" behaviour. • Low cost for government. • Inspires movement beyond minimum compliance.	• Freedom of choice may be contested. • Leaves harm unaddressed if people or organisations do not respond to nudge. • Possible small effect for complex decisions.
Information supply.	Mandatory energy performance certificates.	• Low cost for government. • Direct link between intervention and outcome. • Freedom of choice in how to comply. • Increases informed choice for end-users.	• High enforcement costs. • Cost of producing information may be high for those subject to intervention. • End-users may not understand information provided.
	Voluntary energy performance benchmarking and rating.	• Inspires movement beyond minimum compliance. • Rewards those who move beyond minimum compliance. • Direct link between intervention and outcome.	• Cost of producing information may be high for those subject to intervention. • End-users may not understand information provided. • Intervention may be captured by business interests.

(Source: Author, but note this table takes inspiration from Baldwin et al. (2012, Chapter 7)).

interventions are necessary but not sufficient to accelerate building retrofits at high speed and on a large scale. However, if both coercive and persuasive interventions fall short, is all hope lost for governance interventions for building retrofits? I think not. We do, however, have to think differently about how we use them.

5.5 Embracing diversity: beyond either/or, beyond one-size-fits-all

The main challenge, of course, is that we are running out of time. Despite decades of coercive and persuasive governance initiatives, we are still stuck with a building stock that is vastly energy inefficient and unsustainable (BPIE, 2017; GlobalABC, 2020; IEA, 2020). A way out of the conundrum explored in this chapter – that is, the limits of both coercive and persuasive governance interventions for building retrofits – is to embrace the diversity provided by these, as well as the diversity of the individuals and organisations that they target.

First, governments often approach these interventions as an either/or choice; they seek building retrofits through *either* coercive *or* persuasive interventions. Experience has taught us, however, that there is no single optimal pathway towards achieving building retrofits. These interventions provide us with multiple pathways towards the same end, and some pathways are attractive to some sub-groups whilst others are attractive to others. Governments can take up the role of the "orchestrator" of these interventions and make sure that a broad variety of interventions is attractive to as large a group of property developers, owners, and tenants as possible.

Second, experience has taught us that these individuals and organisations can, at least, be clustered as early movers, the majority, or laggards. Each cluster (or sub-group) has its own wishes and needs to be challenged with different incentives. Governments responsible for systems of building regulation and governance, however, typically do not embrace these differences. They often apply a "one size fits all" strategy for targeting these property developers, owners, and tenants. For example, they may increase bottom-line standards (which will push up laggards, *but not* the majority or the early movers), meet individuals and organisations in the middle with nudges, limited subsidies and tax breaks (which is attractive for the majority, *but not* the early movers or the laggards), or support innovation through voluntary programmes and experiments (which will pull up early movers, *but not* the majority or the laggards).

Embracing diversity, then, implies that, rather than introducing one coercive or persuasive governance intervention to replace or overcome the weakness of another, a mix of interventions should be introduced that all have strengths *and* weaknesses but have a combined strength that is larger than the sum of its parts (and a combined weakness that is smaller than the sum of its parts). Figure 5.1 presents an analytical ideal-type model for such a mix.

In summary, Figure 5.1 illustrates how the full market of early movers, the majority, and laggards can be targeted with a variety of governance interventions that suit each subgroup's individual needs. Early movers can be incentivised by

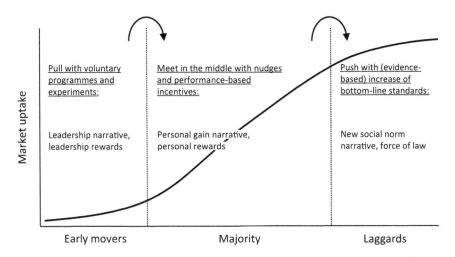

Figure 5.1 Embracing diversity in the regulation and governance of building retrofits
Source: Author

presenting a leadership narrative that stresses the need to move early and the advantages of doing so, and that provides early movers with leadership rewards through voluntary programmes and experiments that include not only monetary and information rewards but also recognition of their early action ("leadership rewards"). The market's majority can be met in the middle by using nudges which help to overcome their biases against retrofitting, and performance-based incentives such as tax-breaks and targeted subsidies. This group can be presented with a narrative of how retrofitting their property will result in personal gain through these personal rewards. Finally, the laggards will ultimately need to be pushed with an increase in the bottom-line standards; these can be introduced within a larger narrative of changing social norms around retrofitting and be backed by the force of law.

What is more important is that Figure 5.1 stresses the need to think about the *full* system of building regulation and governance when seeking to accelerate the scale and speed of building retrofits. Not only can a mix of interventions ensure that the whole market of property developers, owners, and tenants is targeted with tailored interventions, but these interventions can also mutually reinforce each other. Lessons learnt from voluntary programmes and experiments applied by early movers may help to develop nudges and performance-based incentives that are highly attractive for the majority. Lessons learnt from all these interventions will allow evidence-based mandatory building regulation to be developed that "works" for laggards. New bottom-line standards will again be the jumping board for early movers to explore even better solutions through improved voluntary programmes and experiments. Rather than getting stuck in the limitations of either/or choices and one-size-fits-all strategies, such a system may result in an upward spiral of ever-improving building regulation and governance for achieving building retrofits and, more broadly, urban climate mitigation and adaptation, as well as buildings and cities that are healthy to live in a make a positive contribution to our overall well-being.

Notes

1 https://sfgov.org/esip/seismic-retrofit-financing (4 May 2020).
2 www.energy.gov/eere/slsc/property-assessed-clean-energy-programs (4 May 2020).
3 www.building.govt.nz/managing-buildings/managing-earthquake-prone-buildings/how-the-system-works/ (6 May 2020).
4 www.usgbc.org/leed/rating-systems/existing-buildings (12 May 2020).
5 www.breeam.com/discover/technical-standards/breeam-in-use/ (12 May 2020).

References

Baldridge, K., 2012. Disaster resilience: A study of San Francisco's soft-story building problem. *The Urban Lawyer*, 44(2), pp. 465–492.

Baldwin, R., Cave, M. and Lodge, M., 2012. *Understanding Regulation: Theory, Strategy and Practice*, 2nd edition. Oxford University Press, Oxford.

Barber, B., 2013. *If Mayors Ruled the World*. Yale University Press, Newhaven.

BPIE, 2015. *Energy Performance Certificates Across Europe: From Design to Implementation*. Buildings Performance Institute Europe, Brussels.

BPIE, 2017. *97% of Buildings in the EU Need to Be Upgraded*. Buildings Performance Institute Europe, Brussels.

BPIE, 2019. *Future-Proof Buildings for All Europeans: A Guide to Implement the Energy Performance of Buildings Directive*. Buildings Performance Institute Europe, Brussels.

Bulkeley, H., 2010. Cities and the governing of climate change. *Annual Review of Environment and Resources*, 35(1), pp. 229–253.

Carroll, W., 2007. Global cities in the global corporate network. *Environment and Planning A*, 39(10), pp. 2297–2323.

Carson, J., McCulloch, K. and Pooser, D., 2013. Deciding whether to invest in mitigation measures: Evidence from Florida. *Journal of Risk and Insurance*, 80(2), pp. 309–327.

Chakos, A. and Zoback, M. L., 2018. California's San Francisco Bay Area: An epicenter of community resilience. In J. Bohland, J. Harrald and D. Brosnan (Eds.), *The Disaster Resilience Challenge: Transforming Theory to Action* (p. 211). Charles C Thomas, Springfields, IL.

Chhotray, V. and Stoker, G., 2010. *Governance Theory and Practice*. Palgrave, Houndmills.

Cowling, K. and Tomlinson, P., 2005. Globalisation and corporate power. *Contributions to Political Economy*, 24(1), pp. 33–54.

Drahos, P. (Ed.), 2017. *Regulatory Theory*. ANU Press, Canberra.

Egbelakin, T., Wilkinson, S., Ingham, J. and Potangaroa, R., 2017. Incentives and motivators for improving building resilience to earthquake disaster. *Natural Hazards Review*, 18(4), pp. 1–15.

Filippova, O. and Noy, I., 2020. Earthquake-strengthening policy for commercial buildings in small-town New Zealand. *Disasters*, 44(1), pp. 179–204.

GlobalABC, 2020. *Global Status Report of Buildings and Construction*. Global Alliance for Buildings and Construction, Paris.

Hodge, G. A., 2000. *Privatization: An International Review of Performance*. Westview Press, Boulder.

Hoffmann, M., 2011. *Climate Governance at the Crossroads*. Oxford University Press, Oxford.

Hyde, R., 2013. Reviewing benchmarking systems for retrofitting: How can benchmarking be harnessed for the purposes of retrofitting? In R. Hyde, N. Groundout, F. Barra and K. Yang (Eds.), *Sustainable Retrofitting of Commercial Buildings: Warm Climates*. Routledge, London.

IEA, 2020. *Energy Efficiency 2020*. International Energy Agency/OECD, Paris.

Jordan, A. and Huitema, D. (Eds.), 2019. *Polycentricity in Action*. Oxford University Press, Oxford.

Law, A. and Butterworth, N., 2019. Prescription in English fire regulation: Treatment, cure or placebo? *Forensic Engineering*, *172*(2), pp. 61–88.

Leon, R. and Rossberg, J., 2012. Evolution and future of building codes in the USA. *Structural Engineering International*, *22*(2), pp. 265–269.

May, P., 2011. Performance-based regulation. In D. Levi-Faur (Ed.), *Handbook on the Politics of Regulation* (pp. 373–384). Edward Elgar, Cheltenham.

McRea, C., McClure, J., Henrich, L., Leah, C. and Charleson, A., June 2018. Reactions to earthquake hazard: Strengthening commercial buildings and voluntary earthquake safety checks on houses in Wellington, New Zealand. *International Journal of Disaster Risk Reduction*, *28*, pp. 465–474.

Meacham, B., 2016. Sustainability and resiliency objectives in performance building regulations. *Building Research & Information*, *44*, pp. 474–489.

Meacham, B., Bowen, R., Traw, J. and Moore, A., 2005. Performance-based building regulation: Current situation and future needs. *Building Research & Information*, *33*(2), pp. 91–106.

Osborne, D. and Gaebler, T., 1992. *Reinventing Government: How the Entrepreneurial Spirit Is Transforming the Public Sector*. Addison-Wesley Publishers, Reading.

Pérez-Lombard, L., Ortiz, J., González, R. and Maestre, I. R., 2009. A review of benchmarking, rating and labelling concepts within the framework of building energy certification schemes. *Energy and Buildings*, *41*(3), pp. 272–278.

Rhodes, R. A. W., 1996. The new governance: Governing without government. *Political Studies*, *44*(4), pp. 652–667.

Rhodes, R. A. W., 2007. Understanding governance: Ten years on. *Organization Studies*, *28*(8), pp. 1243–1264.

Rogers, E. M., Medina, U., Rivera, M. and Wiley, C., 2005. Complex adaptive systems and the diffusion of innovations. *Innovation Journal*, *10*(3), pp. 1–26.

Romero-Lankao, P., Burch, S. and Hughes, S., 2018. Governance and policy. In C. Rosenzweig, W. Solecki, P. Romero-Lankao, S. Mehrotra, S. Dhakal and S. Ibrahim (Eds.), *Climate Change and Cities: Second Assessment Report of the Urban Climate Change Research Network* (pp. 585–606). Cambridge University Press, Cambridge.

Sanchez, A., van der Heijden, J. and Osmond, P., 2018. The city politics of the urban age: A literature review of urban resilience conceptualisations and policies. *Palgrave Communications*, *4*, article 25.

Sassen, S., 2015. Bringing cities into the global climate framework. In C. Johnson, N. Toly and H. Schroeder (Eds.), *The Urban Climate Challenge* (pp. 24–36). Routledge, London.

Sayce, S. and Hossain, S., 2020. The initial impacts of Minimum Energy Efficiency Standards (MEES) in England. *Journal of Property Investment & Finance*, *38*(5), pp. 435–447.

Sichtermann, J., 2011. Slowing the pace of recovery: Why property assessed clean energy programs risk repeating the mistakes of the recent foreclosure crisis. *Valparadiso University Law Review*, *46*(1), pp. 263–309.

Solecki, W., Rosenzweig, C., Dhakal, S., Roberts, D., Salisu Barau, A., Schutz, S. and Ürge-Vorsatz, D., 27 February 2018. City transformations in a 1.5 °C warmer world. *Nature Climate Change*, 8, pp. 177–181.

Trencher, G., Castan Broto, V., Takagi, T., Springings, Z., Nishida, Y. and Yarime, M., 2016. Innovative policy practices to advance building energy efficiency and retrofitting: Approaches, impacts and challenges in ten C40 cities. *Environmental Science & Policy*, 66(1), pp. 353–365.

Trencher, G. and van der Heijden, J., August 2019. Instrument interactions and relationships in policy mixes: Achieving complementarity in building energy efficiency policies in New York, Sydney and Tokyo. *Energy Research & Social Science*, 54, pp. 34–45.

van der Heijden, J., 2009. *Building Regulatory Enforcement Regimes. Comparative Analysis of Private Sector Involvement in the Enforcement of Public Building Regulations.* IOS Press, Amsterdam.

van der Heijden, J., 2014. *Governance for Urban Sustainability and Resilience: Responding to Climate Change and the Relevance of the Built Environment.* Edward Elgar, Cheltenham.

van der Heijden, J., 2015a. On the potential of voluntary environmental programmes for the built environment: A critical analysis of LEED. *Journal of Housing and the Built Environment*, 30(4), pp. 553–567.

van der Heijden, J., 2015b. Regulatory failures, split-incentives, conflicting interests and a vicious circle of blame: The new environmental governance to the rescue? *Journal of Environmental Planning and Management*, 58(6), pp. 1034–1057.

van der Heijden, J., 2015c. What 'works' in environmental policy-design? Lessons from experiments in the Australian and Dutch building sectors. *Journal of Environmental Policy & Planning*, 17(1), pp. 44–64.

van der Heijden, J., 2017a. Brighter and darker sides of intermediation: Target-oriented and self-interested intermediaries in the regulatory governance of buildings. *Annals of the American Academy of Political and Social Science*, 670(1), pp. 207–224.

van der Heijden, J., 2017b. Eco-financing for low-carbon buildings and cities: Value and limits. *Urban Studies*, 54(12), pp. 2894–2909. doi:10.1177/0042098016655056.

van der Heijden, J., 2017c. *Innovations in Urban Climate Governance: Voluntary Programs for Low Carbon Buildings and Cities.* Cambridge University Press, Cambridge.

van der Heijden, J., 2018. Hybrid governance for low-carbon buildings: Lessons from the United States. In S. Wilkinson, T. Dixon, N. Miller and S. Sayce (Eds.), *Routledge Handbook for Sustainable Real Estate* (pp. 55–69). Routledge, Milton Park.

van der Heijden, J., 2020a. Understanding voluntary program performance: Introducing the diffusion network perspective. *Regulation & Governance*, 14(1), pp. 44–62.

van der Heijden, J., 2020b. Urban climate governance informed by behavioural insights: A commentary and research agenda. *Urban Studies.* doi:10.1177/0042098019864002.

van der Heijden, J., Bulkeley, H. and Certomá, C., 2019. *Urban Climate Politics: Agency and Empowerment.* Cambridge University Press, Cambridge.

van der Heijden, J. and Van Bueren, E., 2013. Regulating sustainable construction in Europe: An inquiry into the European Commission's harmonization attempts. *International Journal of Law in the Built Environment*, 5(1), pp. 5–20.

van der Heijden, J. and Hong, S.-H., 2020. Urban climate governance experimentation in Seoul: Science, politics, or a little of both? *Urban Affairs Review.* doi:10.1177/1078087420911207.

Young, M., Cleary, K., Ricker, B., Taylor, J. and Vaziri, P., May–June 2012. Promoting mitigation in existing building populations using risk assessment models. *Journal of Wind Engineering and Industrial Aerodynamics*, 104–106, pp. 285–292.

6 Financing retrofits

Ursula Hartenbeger, Sarah Sayce and Zsolt Toth

6.1 Introduction

This chapter examines the ways in which financing for retrofit is evolving to assist owner-occupiers and investment owners to gain finance, from either or both public and private financiers to support retrofit work. It concludes that the situation is evolving rapidly, and arguments previously made that funding was not available have now started to be overcome. However, it concludes that more needs to be done and puts forward some recommendations to aid both the speed and quantum of progress.

As argued earlier in the book, buildings are possibly the largest energy consuming sector, responsible for 38% of total global energy-related carbon dioxide (CO_2) emissions (Global Alliance for Buildings and Construction, 2020). This presents a physical – but also financial – challenge if, as targeted by international cooperative agreements, direct building CO_2 emissions are to halve from 2020 by 2030 to get on track for achieving a net zero carbon building stock by 2050. And this target relates only to direct (Scope 1) carbon. The bigger challenge of Scopes 2 and 3[1] presents an even greater challenge. It must not also be forgotten that the ambition of decarbonisation is to provide occupiers with the range of other benefits that flow from such work (Kerr et al., 2017).

Much of the emphasis of this chapter is on Europe. This is because Europe has the oldest building stock of all major developed regions with an estimated three-quarters of buildings considered to be inefficient, while only 0.4%–1.2% (depending on the respective country) of the building stock is renovated each year to improve overall efficiency (European Commission, 2019). To move the whole building stock onto a net zero emission pathway, renovation rates and depths have to be significantly increased and this requires finance, which is the key focus of this chapter. In other regions of the world, where building stock may be newer, this does not guarantee higher standards – or less carbon intensity of construction or use.

Furthermore, the chapter concentrates on finance schemes that are geared towards the residential sectors although many also cover commercial loans and grants. But the emphasis on residential is probably more critical to enhancing the stock. In the commercial sector, not only are more traditional funding options

DOI: 10.1201/9781003023975-8

available for borrowing or raising corporate funding by equity or bond issue but the pressures of ESG (Environmental, Social and Governance) policies and reporting are leading to changed behaviours as considered in Chapter 8. It is primarily within the residential owner-occupied sector that direct stimuli at the asset level are required.

It has become clear that directing substantially more finance and investment towards building retrofits through dedicated policy and market instruments is critical to achieving committed carbon targets. Without the money, retrofit rates will simply not increase sufficiently, even given the potential in many cases for equity funded retrofits to result in capital value or rental increases. While policy tools such as Energy Performance Certificates (EPCs) have contributed to raising awareness about energy efficiency and are now well-established in some jurisdictions, they have not succeeded in sufficiently ramping up renovation rates. Exceptions to this are beginning to appear, such as in the UK and EU, where mandatory EPC levels are linked to the ability to let through setting minimum energy efficiency standards.

Additionally in the EU and UK, there are proposals to set targets to link EPC rating to domestic loan books for lenders. This move could have a negative effort on stimulating retrofits as lenders might focus only on already 'green' stock. The same applies to the majority of public subsidy and grants schemes which have also not yielded the desired uptake, especially by homeowners, resulting in the demise of the initiative.

In addition, the year 2020 saw unprecedented global economic and social challenges in the form of the Covid-19 pandemic, resulting in massive stimulus packages being agreed upon by governments. While the initial focus obviously has to be on protecting health, incomes and businesses, building future capacity and resilience into social and economic systems is recognised by many governments as being equally important. Hence, policy safeguards need to be built into any recovery plans to avoid public money supporting environmentally or socially harmful activities, potentially worsening the current but also any future crises, locking in highly polluting infrastructure or promoting activities that do not contribute to social equity.

However, public subsidised funds can only do so much: private equity and debt funding are key to ensuring a sea-change in retrofit rates. Equally critical to any real measure of success is that retrofits are based on the principles of resilience to physical climate change, not just designed in anticipation of regulatory change; and that the importance of the health and well-being agenda, which came so evidently to the fore as part of the pandemic, is not lost.

6.2 Why are resilient retrofits so low? The funding barriers

In principle, money that could be deployed to kickstart financing of large-scale renovation programmes which incentivise energy efficiency and ESG-orientated retrofits is now more readily available than in the past, in the shape of both private sector and public funds. Despite this, progress remains sluggish. The challenge now, therefore, is not so much the non-existence of available funds or financing but how to successfully channel more investment to projects that can provide a substantial climate change mitigation impact while also supporting changed post-pandemic occupational requirements. Unfortunately, this is not as straightforward as it may sound due to existing sectoral and systemic barriers.

6.2.1 Sectoral fragmentation of whole lifecycle thinking

Three overarching sectoral obstacles to increasing investments in resilient improvements are listed as follows:

6.2.1.1 Sectoral fragmentation and lack of whole lifecycle and circular economy thinking

The construction and real estate sector is traditionally known for its fragmentation. A multitude of stakeholders make up the sectoral life cycle. The problem is that these stakeholders tend to operate in their respective lifecycle stage silos with very limited, if any, inter-stage communication and feedback mechanisms. When designing and developing, rarely is any thought being given to future operation and ease of retrofit as a completely different set of stakeholders holds responsibility for these activities.

6.2.1.2 Real and perceived lack of data

Whether constructing, refurbishing, buying, leasing or occupying real estate assets, any type of investment decision and, ultimately, the effectiveness of policy instruments and the successful articulation of the business case for investing in resilient retrofits depend on the existence of whole lifecycle data to build environment value chain stakeholders.

At present, systematic data collection and management along a building's life cycle is still more an exception than a rule. There are a plethora of reasons why the construction and real estate sector, unlike other industries, is still struggling with data capture and its subsequent management. Sectoral fragmentation and the associated lack of holistic, whole lifecycle thinking and practice is definitely one of them as it results in a multitude of disconnected data and information owners with very different – often vested – interests as to how to use and share the information (see Figure 6.1).

However, while in practice there may remain real data gaps, the perception of lacking data often arises due to an actual lack of a place to store the information. Data that were there at some point in time, typically at the point of planning and design or during transactions, are getting lost, scattered across different locations or captured and processed in such a way that it is no longer useable when needed to support building work. This in turn can undermine the ability to raise finance,

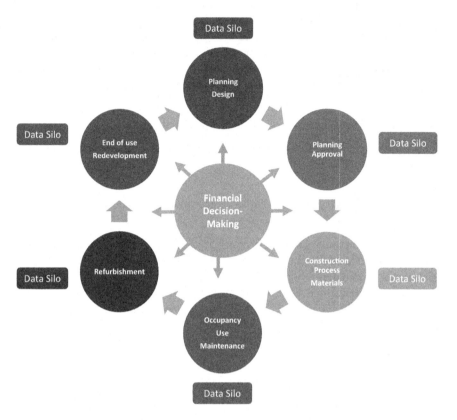

Figure 6.1 Sectoral data fragmentation as a barrier to stepping up investment in renovation

Source: (adapted from Hartenberger and Lorenz, 2018)

particularly preferential finance, as lending institutions are unable to match building performance data with financial data.

To date, there is no universal, standardised system or protocol in place that would easily facilitate access, storage, update and transfer of building-related data and information in a standardised format to assist financing the value chain. The development of building 'passports' (see Section 6.4) may be influential in helping to overcome some of the data barriers.

6.2.1.3 *Systemic and structural weaknesses in relation to existing retrofit financing programmes*

There is generally no shortage of public financing and incentives and grant schemes; in addition, in a low interest rate environment, the current cost of borrowing money is low. Yet the uptake, with some exceptions, is also perceived to be low. Two possible reasons for this are as follows:

a)   Some incentive schemes are simply perceived as too onerous and complicated in terms of the bureaucratic burden they are putting on the applicants who may also find it difficult to navigate the various schemes and programmes;

b)   A misalignment of public incentives too often means that money and time are wasted with short-lived schemes that focus on one single measure rather than taking a holistic view that would encourage a more comprehensive renovation of the building and ensure delivery of overall optimised performance and policy targets. In other words, it is important to fully align any public financing with climate pathways while developing or adapting schemes that make it easy for interested parties to access public finance.

6.2.2 Barriers at investor/financing institution level

In terms of private sector funding of renovation there are currently five main areas of concern:

6.2.2.1 Complex and risky

Financing retrofit and refurbishment projects can be seen as more complex, challenging and riskier than building new, not only by asset owners or prospective buyers but also by financing institutions. Especially for financing small-scale projects, many banks may prefer to finance via personal loans with less favourable interest rates rather than issuing mortgages. This could be due to the cost of obtaining the requisite valuations to support the loan security but also may be due to the fact that banks are simply more used to issuing consumer loans than secured improver loans (normally referred to as green mortgages) which are a relatively recent product. Banks are also more knowledgeable about structuring consumer loans for capital adequacy requirements. In practice this means: when purchasing a low-performing asset with a view to renovate and upgrade sustainability performance, typically a mortgage is issued for both the value of the asset plus the cost of the work, regardless of whether or not the retrofit will significantly improve the resilience score of the building. Sometimes the loan may be made with a retention as to the amount of the advance until the work is 'signed off' as complete, but this presents the borrower with the need to either fund the works from equity or to borrow specifically against the costs at an inevitably increased cost of financing. This may be particularly challenging in the domestic sector where loan to value levels tend to be high and purchasers simply may not have other funds. The result is that potential works remain unfulfilled.

6.2.2.2 A lack of internal protocols

Private retrofit finance and secured lending are hampered both by banks' internal protocols and risk assessment and risk management processes that tend to focus on the credit worthiness of the borrower rather than the project to be financially supported. In the case of commercial schemes, this means that the lending is often determined at the corporate, rather than asset level. This means that, while the borrower's overall policies towards ESG may be considered by the lender, the actual retrofit scheme is not linked directly to the loan awarded. In the domestic sector, lending is linked to the asset level combined with the income or equity profile of the borrower. Typically, the valuations, which support the asset level loan, are not at a sufficiently high fee level to enable the lender to receive sufficiently fine-grained advice to tie in the loan to any increase in climate resilience.

6.2.2.3 Still a new field for private financiers

Unlike the investment community, financing institutions and mortgage lenders have only recently started to engage in earnest with green financing and the development of dedicated products. This results in significant experience, skills and knowledge gaps regarding how lenders instruct their advisors, such as valuers undertaking valuations for mortgage lending purposes and the response by the advisors.

6.2.2.4 Regulatory controls on lending

The international regulatory framework guiding today's banking system (The Basel Framework[2]) is driven by the overriding objective of ensuring financial stability by increasing both the quantity and quality of regulatory capital and liquidity. Currently, this does not fully acknowledge the systemic risks posed by climate change.

In Europe, at present the implementation of capital adequacy and transparency requirements through revised Capital Requirements Regulations (CRR 2) and Directives (CRD 5) may actually hinder the facilitation of preferential financing conditions for sustainable projects by banks – including supporting building retrofits. However, the recently published EU Strategy for Financing the Transition to a Sustainable Economy[3] is set to address these barriers and in other geographies, such as the US and Australia, the Basel Framework is in various stages of adoption, and with it, varying implications for the approach of national banks towards credit risk in relation to property lending.

Overall, though, the increased concern for loan liquidity and credit risk may underscore caution to lending for what could be perceived as risky retrofit endeavours. Over time, obligations on lenders to maintain certain liquidity levels mean they have an increased need to understand future risk of their property loan books. As climate risk increases, non-resilient stock may become less easy to sell – that is, increasing in illiquidity (Clayton et al., 2021).

6.2.2.5 A definitional issue

Finally, in the past, the lack of standardisation and the lack of an agreed definition and parameters for what constitutes a sustainable real estate investment, in general, and a sustainable renovation project, in particular, has proved to be a major issue. This applies to both the public and private sectors. There is no shortage of commercially driven sets of indicators but this does, in fact, undermine the creation of a common language that would help to build confidence in such investments by the investor and finance community. The lack of standardised underwriting processes, certifications and technical knowledge results in high transaction and operation costs, meaning financial institutions are generally reluctant to lend to energy efficiency and other retrofit projects.

To illustrate the size of the challenges mentioned earlier, the example of work by the Energy Efficiency Financial Institutions Group (EEFIG) is cited. Their 2015 report (EEFIG, 2015) identified key financing issues as set out in Table 6.1.

Table 6.1 Overview of barriers to long-term building retrofits

Themes	*Barriers and needs*
Market demand	Market distortions (e.g. excessive subsidisation of energy prices) can prevent building owners receiving accurate price signals that reflect the true marginal cost of the energy use
	Split incentives between owners and tenants
	Lack of regulatory stability and consistency hampering confidence in long-term investment decisions
Financing	High capital costs, long payback times and dispersed benefits and financial returns
	Deep renovations are technically complex, risky and involve high transaction costs
	Aggregation of similar energy efficiency investments into bundles
Information	Quality and availability of EPCs to prepare informed investment decisions
	Lack of common data protocols for collecting and disclosing of data related to energy efficiency investments
Regulatory/Institutional	Capital requirements are not aligned with environmental risks
	National long-term renovation strategies adequately identify the funding streams required especially for the worst performing building stock
	Uncertainties of predicted energy cost savings and the insufficient evidence base linking energy efficiency to probability of default
	Lack of common underwriting processes for debt and equity investments
	Lack of standardised evaluation methods for measuring and verification of risks and benefits
Technical	Insufficient capacity to develop, plan, implement and monitor deep renovation packages
	Systematic ex-post evaluation is still not mainstream

(Source: adapted from EEFIG and Lohse and Zhivoc, 2019).

Progress in developing channels for financing energy efficiency and resilient retrofits, barriers remain requiring concerted stakeholder actions, a point underscored by Bertoldi et al. (2021).

6.3 From barriers to solutions: examples of funding of resilient retrofit projects

A range of 'smart' financial schemes that, almost always, owe their origin to supportive policy measures aim to overcome the financial barriers. Examples of such schemes may involve finance to upskill building professionals, technical assistance to develop projects and awareness raising campaigns to highlight the multiple benefits of retrofit rather than redevelopment.

Public sector-led initiatives are usually designed to overcome the initial barriers of financing energy efficiency projects and include grants, subsidies, debt financing, tax incentives and guarantees. These are instruments which aim to trigger additional private funding as public financing alone is not enough to meet the required investment volume.

Public-led schemes have often led the way to ensure a supply of funds. Whilst private sector schemes are being developed to finance market transformation. The latter will be critical, given the pressures on public sector finances resulting from Covid. Such schemes include green mortgages, green bonds, energy service companies (ESCOs) and on-bill financing.

To date, such schemes, not always successful, are focused almost exclusively on improvements to energy efficiency. There remains much to do to extend financial packages to cover the full range of asset resilience improvement measures, including decarbonisation, flood and storm protection and retrofitting design to reduce heating and cooling needs, depending on geographies.

Figure 6.2 provides a schematic overview of existing schemes aimed at increasing the rate of energy efficiency retrofits.

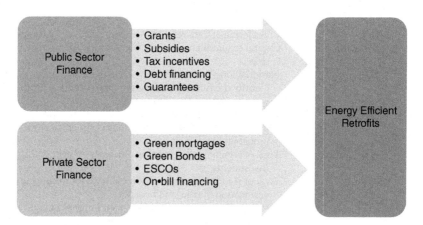

Figure 6.2 Public and private sector financing schemes for funding energy-efficient retrofits

6.3.1 Public sector-led schemes

Not only did the 2015 Paris Climate Agreement (United Nations, 2015) create an international agreement with regard to global CO_2 reduction targets, but it also recognised the urgent need for governments to align financial flows with a pathway towards low-carbon and climate-resilient buildings in order to achieve these targets.

The scale of the challenge is elaborated in other chapters; here a number of examples of publicly led examples to fund retrofitting are provided. It is acknowledged that these present only a small sample of schemes that have proliferated around the world in response to the growing recognition of the need to retrofit existing buildings to support carbon reduction policies. At this point, it is noted that all the schemes discussed have energy efficiency or/and decarbonisation as their objective; little has so far been achieved towards the wider aspects of climate resilience and this is commented on in the final section of the chapter.

6.3.1.1 Example 1: the EU sustainable finance taxonomy: creating a common language for investment in renovation

The EU Action Plan for Sustainable Finance in March 2018[4] included the development of a Taxonomy[5] for determining what constitutes a green/sustainable investment. This Taxonomy aims to create a common understanding as a tool to navigate stakeholders towards low-carbon, resilient and resource-efficient investments – including building retrofit investments – by setting performance thresholds to focus money towards at least one of six environmental objectives (climate change mitigation and adaptation, water, pollution, circular economy and ecosystems) while allowing no significant harm (DNSH) to the other five.

6.3.1.2 Example 2: Germany KfW loans and grants

The residential energy efficiency schemes run by the German State Investment Bank KfW are, arguably, among the longest and most successful schemes issuing both grants and loans towards renovation and construction of energy-efficient homes. While funding many other forms of innovation, including wind farms, the home improvement grants have, since the 1990s, provided a stable platform and, by 2010, had supported high energy retrofits to in excess of 9 million pre-1979 housing units (Schröder et al., 2011). Their prerequisite of having an energy efficiency specialist assessment before any application is made, together with monitoring of the works, is critical to achieving high standards of construction. Although generally claimed to be a leading scheme in renovations, Rosenow and Galvin (2013), in evaluating the schemes, conclude that the carbon savings may be over-calculated and that many of the works would have taken place even without the scheme. Nonetheless, the longevity of the scheme does provide confidence to those seeking to make improvements.

6.3.1.3 Example 3: Building Retrofit Energy Efficiency Financing (BREEF), Singapore

The Building Retrofit Energy Efficiency (BREEF) scheme 16,[6] facilitated by Singapore's Building and Construction Authority (BCA) in collaboration with participating financial institutions, enables financing of up to 90% for upfront costs of energy retrofits of existing commercial or residential buildings, subject to a total loan cap, through an energy performance contract arrangement. The scheme covers the cost of equipment, installation and also any related professional fees. The retrofit has to achieve as a minimum Building Certification Award (BCA) of Green Mark Certification 12 standard, which is maintained throughout the loan tenure. However attractive this scheme may appear, the loan is interest-bearing and is billed as of limited life. This life limitation may well adversely impact its overall success.

6.3.1.4 Example 4: Clean Energy Corporation, Australia

Within Australia, there exist a wide range of incentives to upgrade for low carbon solutions, both new build and, pertinent here, for upgrading existing properties. However, the schemes vary from state to state and many are loans, rather than grants, and may be based on commercial return criteria.

To that extent, although the government does run some schemes directly, one of the biggest finance facilitators is the Clean Energy Finance Corp (CEFC), whose role is to increase the flow of private sector money into the market, using its powers of communication and customer reach; it also acts as a co-funder in some cases. However, the objective of the government is to, in effect, support and encourage the commercialisation of a move towards the adoption of renewable energy – rather than to simply pay the upfront costs of investment. Therefore, the over-arching principle would be to stimulate demand and ease the capital flows rather than, in most cases, subsidise transition.

6.3.1.5 Example 5: Property Assessed Clean Energy (PACE), the US

The PACE[7] scheme, run by the US Department of Energy's Office of Energy Efficiency and Renewable Energy (EERE) is a programme for financing energy efficiency and renewable energy improvements for privately owned residential (R-PACE) and commercial (C-PACE) properties. The R-PACE scheme is currently available in three US states. The C-PACE programmes exist in more than 35 states, regions and local governments.

The PACE programmes allow property owners to finance the up-front cost of energy efficiency or other eligible improvements and pay the costs back over a period of typically 10–20 years through property assessments, which are secured by the property itself rather than the individual and paid as an addition to the owners' property tax bills, thereby providing a mechanism which is, essentially, easy for the end-user.

Because the PACE assessment is a debt of property, meaning the debt is tied to the property and not to the property owner, the repayment obligation may transfer with property ownership if the buyer agrees to assume the PACE obligation and the new first mortgage holder allows the PACE obligation to remain on the property. This possibility for transferability can help address a key disincentive to investing in energy improvements because many property owners tend to be reluctant to undertake improvements if they think they may not stay in the property long enough for the resulting savings to cover the initial upfront costs. However, it could have an impact at time of sale.

Finally, PACE allows for secure financing of comprehensive projects over a lengthy time period, helping to make projects cash flow positive. By being linked to the property tax bill, the high security nature of the loan repayments due to it may also result in low-interest rates. For these reasons, and in the light of empirical work in California, Ameli et al. (2017) conclude that the scheme has been highly cost-effective in stimulating retrofit technologies.

6.3.1.6 Example 6: mixed results, The UK

The UK has introduced many schemes aimed at home improvement, from the Feed-in Tariff (FiT), which succeeded in stimulating installation of solar panels by guaranteeing payment of power returned to the grid, to the renewable heat incentive (RHI), which provided financial assistance towards the installation of, for example, biomass boilers and heat pumps. Both aimed to stimulate the private sector market for existing properties. The FiT has now ceased, with the RHI to close in 2022. Arguably both achieved some success in stimulating development but overall had limited impact.

The most ambitious scheme was the Green Deal, introduced following the Energy Act (2011). This pay-as-you-save scheme was aimed at homeowners who would commission work from private accredited contractors on the basis that the installation of the improvement, typically solar panels, would involve no upfront cost with repayment being linked to a maximum of energy savings achieved and the debt secured against the property. However, following criticism of the policy design, the unattractiveness and complication of the scheme and poor messaging (National Audit Office, 2016; Rosenow & Eyre, 2016), it was withdrawn, with a residue scheme passing to the private sector.

More positively, a recent Welsh government scheme, the Optimised Retrofit programme, offers real potential for a model moving forward, as it is sensitive to individual building typologies and relates to a range of measures. Aimed initially at the social-rented housing sector, the scheme promotes a combination of building fabric improvements, the installation of low and zero-carbon features and smart monitoring controls. Key to the scheme is the development of in-depth whole house survey, recorded digitally, that produces an individualised, alternative 'pathways to zero', which allows for works to be undertaken incrementally mapped to the individual owner's needs for flexibility and sequencing.[8] While funded by the government, the scheme is designed to apply to all tenures; the survey is key and allows owners to choose different routes to either achieve decarbonisation over a short timescale or a slightly longer one which harmonises with decarbonisation of the national electricity grid.

In summary, the key messages for maximising take-up deriving from all these public-sector led schemes can be summarised as:

- the need for consistency and longevity; without it they will fail;
- full commitment by policymakers who have the power to influence positively;
- stimulus by government for private sector money to be devoted to retrofit; and
- ease of use and transparency of schemes.

6.3.2 *Private sector-led schemes*

Even though public sector finance may help to support transformation, stimulation of the provision of private sector finance is critical. Private finance can be more agile and better placed to both change and support increased occupational and investment demand for resilient buildings. Such finance is focused in two ways: provision of funds for buildings which already meet resilient, low carbon standards and, presenting greater challenge, finance to retrofit non-resilient buildings. Examples of private sector finance initiatives are provided as follows.

6.3.2.1 *Example 1: green mortgages*

Green mortgages are preferential loans with financing conditions applied to properties which either meet certain environmental (normally energy) standards, or to finance improvement to the building's energy performance. Most are aimed at the domestic market. Globally, the US has been developing Energy-Efficient Mortgage (EEM) loans from as early as 1980 by adopting special underwriting guidelines to take into account energy efficiency in the mortgage underwriting process. However, for most of the global finance sector they are still at the innovation stage and tend to be loans against buildings that are deemed 'green' – not the retrofit market. While this approach could incentivise building improvements, their usefulness in providing the required finance is diminished if the scope is so restrictive.

Table 6.2 lists examples of selected green mortgage schemes operational in selected countries.

In Europe, the Energy Efficient Mortgages Initiative (EEMI)[9] set up a standardised approach and common definition on the eligibility criteria for assets and projects that can be used for issuing new green loans or to tag existing assets in banks' portfolios (Toth et al., 2018). The framework is complemented by a valuation checklist and a common data reporting and transparency template (Hartenberger et al., 2019) to improve empirical evidence base and demonstrate the de-risking potential of energy efficiency features of buildings. Furthermore, the Energy Efficient Mortgage Label provides additional reassurance for investors, regulators and other market participants and helps to facilitate securitisation and packaging of

Table 6.2 Examples of green mortgage schemes

Initiative	Key features	Website
Nationwide Building Society Green Additional Borrowing Mortgage (UK)	⇒ Choice of either a two or a five year fixed rate product at a discounted initial interest rate compared to standard additional borrowing product range ⇒ No product fees ⇒ Borrowing range between £5,000 and £25,000, dependent on individual circumstances ⇒ Borrowing potential of up to 85% Loan to Value ⇒ Applicant needs to be an existing member with a Nationwide mortgage ⇒ At least 50% of additional borrowing needs to be spent on energy-efficient home improvements, e.g. air source heat pump, cavity wall insulation, double glazing/replacement windows, electric car charging point, ground source heat pumps, loft insulation, small-scale wind turbines, tanks and pipes insulation	www.nationwide.co.uk/products/ mortgages/borrowing-more/green-additional-borrowing
TRIODOS (Belgium)	⇒ Improvement of the energy performance of the home on the basis of an energy performance certificate (PEB) ⇒ The higher the percentage of improvement after the renovation works, the higher the reduction granted (between 0.90% and 0.50%)	www.triodos.be/fr/credit-habitation
Regional Australia Bank Sustainable Home Loan (Australia)	⇒ To be eligible for the Sustainable Home Loan, the home first needs to pass the current minimum environmental standards required by the relevant state or territory authority ⇒ Additionally, evidence needs to be provided that the building has features from two separate list: at least one feature from 'List A': greywater treatment system, solar power system, wind turbine, micro hydro system, double glazed windows or better; and at least two features from 'List B': solar hot water, rainstorm/ water tank, 5 star+ Gas or Electric Heating, external awnings	www.regionalaustraliabank.com.au/ personal/products/home-loans/ sustainable-home-loan
Monmouthshire Building Society	⇒ Funded as part of the UK's VALUER project, this is the first mortgage product to embrace energy costs within their calculation of repayment affordability, thus rewarding energy efficiency at both the point of purchase or to fund improvements ⇒ Trialling started in 2021, with the intention that the pilot will provide a model for further rollouts without being restricted to particular technologies	www.seroprojects.com/wp-content/ uploads/2020/06/290620_ VALUER_Launch_Press-release_ FINAL.pdf

mortgage loans into green bonds. The latter can be an important strategy to refi-
nance these loans and expand the green lending capacity of banks.

Green mortgages may not solve the retrofit financing problem alone, but closer
cooperation between banks and private homeowners has great potential to drive
demand. Energy efficient buildings are more comfortable for occupants and
cheaper to run in terms of utility bills, leading to higher occupier satisfaction. The
value proposition of green mortgages comes from linking these benefits to property
values and financial risk: making dwellings more energy efficient could result in
lower risk of mortgage default, increased risk mitigation capacity and could help to
future proof portfolios against value decline. The reasoning is intuitive, however,
the hard evidence to prove the case is still emerging.

6.3.2.2 Example 2: green bonds

The first green bond issuance, made by the World Bank in 2007–2008, has led to
an important initiative to assist decarbonisation products, and multilateral development banks started to commit to financing 'green' projects. The movement
spread to the private sector in 2013–2014.

Green bonds help mobilise green investments by offering investors an informed,
explicit decision to invest in green projects, including retrofit, and are widely projected to place a critical part in financing decarbonisation and meeting the wider
Sustainable Development Goals. However, they are not without their challenges
(e.g. see Deschryver & De Mariz, 2020; Lashitew, 2021). Projects financed by green
bonds are mainly within renewable energy, energy efficiency, low carbon transport,
sustainable water, and waste and pollution. They offer an alternative to bank loans
and equity financing and enable long-term financing for large-scale projects.

The US is estimated to issue the largest number and value of green bonds,[10] but
the movement is growing globally, with the European green bonds market having
increased an estimated sevenfold since 2015 (Belloni et al., 2020). It has received
a further boost as a response to the Covid-19 economic recovery programme, with
governments' pledges to build back to greener standards.

The challenges discussed earlier in relation to public sector green financing
are the same in this area: lack of green project pipeline, aggregation mechanisms
for green projects, a definition and framework for green bonds, information and
market knowledge and a clear risk profile for green investment. To increase trust
and improve market standards, the Climate Bond Initiative, an investor-focused,
international organisation, has introduced a labelling scheme to certify green
bonds, which as Rose (2018) points out, may need additional steps to verify the
'climate' component, especially in the light of accusations of greenwashing (see,
e.g. Choi, 2020). In response, Europe is also currently exploring the possibility of
a legislative initiative for an EU Green Bond Standard to ensure such control.[11]

6.3.2.3 Example 3: on-bill financing

On-bill financing schemes (OBS) are tied to the utility service and rely on the
utility bills to repay the initial investment. They have been used for over past

30 years in the US and Canada and have encountered some success in overcoming the reluctance of financial institutions to finance small-scale investments. The full replication potential of the on-bill approach is yet to be explored in Europe, with the most widely trailed scheme, the UK's Green Deal, having failed (see Section 6.2).

Within the US and Canada different types of on-bill financing schemes exist, depending on the party that provides the loan or how the efficiency improvement is paid for. These range from on-bill financing (OBF), where the utility company represents the lending party; on-bill repayment (OBR), where the lender is a third party, thereby leveraging private finance; to tariffed on-bill (TOB), whereby the energy efficiency measures are financed through a new tariff offered by the utility provider (e.g. those offered by the American Council for an Energy-Efficient Economy). In reviewing such schemes, Bell et al. (2011) concluded that flexibility and risk-sharing were critical to success; if all the risk falls to the bill-payer, the scheme may fail.

6.3.2.4 Example 4: Energy Service Companies (ESCOs)

ESCOs deliver energy savings by removing the upfront costs of the investment. The investment is repaid through the energy savings achieved through building renovation measures. ESCOs provide an energy savings guarantee or shared savings which shift the performance risk from the end-user to the energy service company. ESCOs operate under different business models, such as energy performance contracting or energy supply contracting. Unlike traditional energy suppliers, they can provide or facilitate financing through a bank loan. Combined with grants and economic recovery funds, ESCOs can be a viable financing mechanism and key instrument driving the energy transition.

Though many ESCO models are currently in use, the approach is not yet standardised and numerous challenges persist. Overall, there is limited awareness of the concept leading to a lack of trust by financial institutions and users. The market uptake is dependent on a well-functioning energy service market and enabling legal framework including a common ESCO definition, certification and standards. Additionally, the main promise of the ESCO model of removing the credit and performance risks as well as offering turnkey renovation solutions requires effective collaboration across the many financial and building sector stakeholders involved.

6.4 Support tools and mechanisms for public and private sector schemes with data

Finance relies on access to good data; this is not always available. Sometimes it is because the data are not collected, but frequently it is because it is not available to decision makers, notably investors, lenders and their advisors. Investment and financing decisions are generally based on trust and confidence, which in turn rely on the availability, transparency and robustness of underlying data and information. Decisions regarding the financing of energy efficient retrofits are no exception. Therefore data are prerequisites to promoting retrofit finance. Two types of initiatives are now being introduced to support data availability.

6.4.1 Example 1: building passports

As Section 6.2 illustrated, data storage is a core issue in stepping up investment in resilient retrofits; lack of adequate data storage option was found by Hartenberger and Lorenz (2018) the single most prominent barrier to bringing market participants' 'data house' in order.

Building Passports, also referred to as Digital Building Logbooks or Building Files, aim to address this issue by providing a whole lifecycle data and information repository of building-related information covering among other aspects, design, physical building characteristics, building operation and management and building materials and equipment throughout the life cycle. The passports are intended to provide a living 'one-stop-shop' containing a mix of traceable static 'as built' and dynamic data and information, and it is envisaged they will be owned and managed by the building owner, who in turn will be able to grant access to the information to third-party stakeholders such as financing institutions for the purpose of mortgage lending (Volt et al., 2020).

One of the challenges to the successful implementation of Building Passports to support financial aid is the plethora of initiatives (as an example, see Sesana et al., 2021). Success in achieving standardisation is deemed critical to facilitating finance (Green Finance Institute, 2021). According to Adisom et al. (2021), much further work is required throughout the supply chain before a Building Passport system can be realised.

6.4.2 Example 2: provision of technical assistance and one-stop-shop approaches

Although public funds and private financing schemes for building retrofits are available, many are concentrated in the residential sector, with the assumption that non-domestic owners will have more access to corporate debt and equity channels. Good communication and high levels of trust and awareness are key for schemes to work in the residential sector. While these are addressed in more detail in Chapter 5, the schemes listed earlier have often suffered from poor take up due to lack of awareness, high transaction costs, lack of standardised procedures and resources to develop viable projects (European Investment Bank, 2020). To help finance for building resilience improvements become more widely taken up and renovation rates increased, significant initial technical assistance and capacity building are needed to define and support ongoing development of project pipelines.

Technical assistance and one-stop-shops (OSS) come in various forms. Project development assistance facilities usually involve training and capacity building of public authorities, financial sector and built environment stakeholders to systematically integrate climate action in their investment plans. They are aimed to build top-down support, including risk assessment, legal advice and marketing towards homeowners and investors and facilitation (Alam et al., 2019; Panteli et al., 2020).

One-stop-shops are advisory tools that can reduce barriers and transaction costs by providing comprehensive information to homeowners on renovation packages, including financial support schemes, technical solutions and implementation support (see, e.g. Cicmanova et al., 2020). One-stop-shops function as a marketplace bringing together public authorities, lenders, energy and construction companies who provide customer reassurance, increase confidence in savings, reduce transaction costs and streamline the financing process.

6.5 Discussion

This chapter has considered many financial schemes, mainly public-sector supported or led, to support retrofitting buildings. Most have been geared primarily towards increasing the volume of energy retrofits within the residential sector. This is understandable given the age and poor resilience of so many domestic buildings.

6.5.1 Incentivising projects: the finance case is only part of the story

Within the owner-occupied sector, the need for climate resilience and energy efficiency is not usually a major concern for homeowners; instead safety, health, comfort and other factors including status, may play a more important role in decision-making (see, e.g. Wilkinson & Sayce, 2019). Therefore, to encourage their participation, financial support requires to be well communicated, easy to obtain, trusted and accompanied by support mechanisms; it also needs to be at preferential rates: a challenge when interest rates are low. If it is not, a litany of abandoned initiatives tells us that take-up will probably be poor.

For these reasons, many attempts have been made to prove a financial case for owners to engage based on the premise that the value of their capital asset (i.e. their home) will support the decision to upgrade. Even though there is some evidence that there is a link between EPC ratings and levels of capital values (see, e.g. Hyland et al., 2013; Högberg, 2013; Fuerst et al., 2015) and rental values (Fuerst et al., 2020), the evidence that upgrade work may lead to increased capital values which exceed the costs involved is not proven. Indeed, Purple Market Research (2019), acting on behalf of the UK government, found in a pilot study of public sector dwellings that installing external wall insulation, while improving the aesthetic, showed limited value increases, well below the cost of the work. Thus, the role of finance remains critical for the domestic sector, especially as the quality of new-builds which are moving towards zero-carbon standards, start to drive purchaser expectations. For this reason, finance which is linked to running cost savings[12] rather than uncertain capital or rental increases, may prove more powerful moving forward for owner-occupiers but will not support investment owners.

For commercial owner-occupiers and commercial investors, the motivation to improve may not require the same level of financial incentive: the rise of ESG and the gradual introduction of requirements to provide carbon disclosure are creating a groundswell of activity. According to the OECD, the volume of ESG investments and assets has risen substantially since 2010 (Boffo & Patalano, 2020).

Further, there is heightened interest within the investment and finance communities to commit to so-called 'green' or resilient real estate assets, including renovation projects, although evidence on this is less secure; indeed, the introduction of regulation regarding the inability to continue to let buildings failing minimum energy standards may be more powerful than direct finance (see, e.g. McAllister & Nase, 2019; Sayce & Hossain, 2020); in other geographies, some also consider mandating may be required (see, e.g. Wrigley & Crawford, 2017).

However, these two elements, finance and mandated standards are likely to be a way forward, as lenders seek to show their own ESG credentials through their loan book. Realising that sustainable finance can have an impact on financial stability and market discipline, in both the UK and EU, regulators are considering disclosure requirements on the extent their loan books align with climate neutrality and resilience (BEIS, 2020; EBA, 2021). If introduced, this would make a very powerful connection between energy performance and the ability to raise finance.

6.5.2 A collaborative need for data improvements

The chapter has shown that obtaining finance is linked to the ability to provide funders with appropriate data. Currently, as the embryonic state of Building Passports and limited data for existing buildings show, there is a large support role for built environment professionals to play to collaborate with each other and their client base to facilitate the production of underlying data. Figure 6.3 presents a

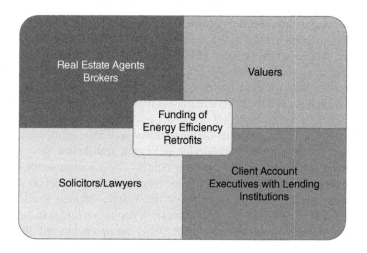

Figure 6.3 Key stakeholders in steering clients towards energy efficiency retrofits

schematic of the key advisers who can help create the required datasets to put together funding requirements.

All these players can play a role in ensuring that their clients' decisions are informed of potential grants and loan schemes and of the changing stance of lenders where appropriate. But to really build the business case required to obtain funding for retrofits requires a radical rethinking of roles and responsibilities within the community of built environment professionals and associated stakeholders, including moving away from traditional silo-thinking towards a knowledge exchange-based multi-disciplinary mentality.

6.5.3 From energy and carbon to full climate resilience: the missing link

This chapter has reviewed a wide range of finance options for retrofitting property to help drive energy efficiency and decarbonisation. It has made the case that the money is there; indeed, from being primarily public-sector initiative driven, increasingly the role of private funding through many routes, notably the bond market, and a rapid increase of preferential secured lending has changed the financial options for retrofit. While some public-sector money is available through grants, preferential loan rates are the more common route. A mixed public-private retrofit financial economy has emerged; at best, these can work together with public sector 'seed corn' money (at least temporarily) reducing the risks for private sector innovation.[13]

The chapter has not addressed funding for infrastructure projects which are increasingly enabling a transition to a fossil-free environment, through 'green grid' electricity, almost without the ultimate electricity consumer knowing. Infrastructure funding is a critical part of the transition journey, but the scope had to be drawn somewhere. However, possibly the bigger omission is the lack of discussion on the funding for true climate resilience measures, that is, for protection against flood, storm, drought and heat.

At the time of writing, the emphasis on developing such resilience lies far more at the government level and the terms of insurability. However, this situation will change as climate events become more widespread, and the impacts on real estate values are likely to increase. As the risk of climate change rises, the channels of insurance and finance will begin to be critical to protect asset values. Currently, these channels remain largely unclear (Clayton et al., 2021), and the investment and finance sectors' response is only just beginning.

6.5.4 Conclusions

The key message of this chapter for both industry players and governments is that for the ambitions of the Paris Accord to be met, finance for decarbonisation and wider building resilience needs to become not just available but transparent and easily accessible. Money is there, but as economies rebuild post-Covid, pressures on the public purse will be extreme. Therefore, the role of the private sector to facilitate becomes ever more critical. But as Bertoldi et al. (2021) concluded in their review of European financial instruments, there is no 'silver' bullet: the very diversity and number of stakeholder interests and the complex nature of the sector preclude this.

This should not be taken as a reason not to innovate; green mortgages, currently focused heavily on stimulating demand for energy efficient buildings need to evolve further to support the whole scope of resilience retrofitting. There are signs that this is happening; however, it needs support and encouragement at state and country level to reduce risks of entering new markets; it also requires transparent data and a sustained business case that such expenditure, even if it will not demonstrably produce quick returns on investment, will act as a counterpoint to building value depreciation.

Notes

1 Carbon emissions, as detailed in Chapter 1, are grouped into scope 1 – directly produced by a company and its buildings; scope 2 covers the carbon emitted by, for example, the fuel used in the building 1 and scope 3 covers all other associated emissions – such as through the supply chain, travel to the building, etc.
2 The Basel Committee on Banking Supervision (BCBS) is the primary global standard setter for the prudential regulation of banks; in particular they set the standards for risk-based capital requirements. (See www.bis.org/basel_framework/)
3 https://ec.europa.eu/finance/docs/law/210704-communication-sustainable-finance-strategy-annex_en.pdf
4 https://ec.europa.eu/info/publications/sustainable-finance-renewed-strategy_en
5 https://ec.europa.eu/info/business-economy-euro/banking-and-finance/sustainable-finance/eu-taxonomy-sustainable-activities_en
6 See: Building Retrofit Energy Efficiency (BREEF) scheme
7 See: www.pacenation.org/what-is-pace/
8 www.optimised-retrofit.wales/
9 See: https://energyefficientmortgages.eu
10 www.institutionalassetmanager.co.uk/2021/01/26/294959/green-bond-issuance-track-almost-double-2021-market-estimates-suggest
11 See: https://ec.europa.eu/info/business-economy-euro/banking-and-finance/sustainable-finance/eu-green-bond-standard_en
12 See for example the VALUER project, listed above, may be more powerful.
13 An example of this is the government sponsored Optimised Retrofit programme, detailed at section 6.5.3.

References

Adisorn, T., Tholen, L. and Götz, T., 2021. Towards a digital product passport fit for contributing to a circular economy. *Energies*, *14*(8), p. 2289. https://doi.org/10.3390/en14082289.

Alam, M., Zou, P. X., Stewart, R. A., Bertone, E., Sahin, O., Buntine, C. and Marshall, C., 2019. Government championed strategies to overcome the barriers to public building energy efficiency retrofit projects. *Sustainable Cities and Society*, *44*, pp. 56–69. https://doi.org/10.1016/j.scs.2018.09.022.

Ameli, N., Pisu, M. and Kammen, D. M., 2017. Can the US keep the PACE? A natural experiment in accelerating the growth of solar electricity. *Applied Energy*, *191*, pp. 163–169. http://dx.doi.org/10.1016/j.apenergy.2017.01.037 0306–2619.

BEIS (Department of Business, Energy and Industrial Strategy), 2020. Improving home energy performance through lenders consultation on setting requirements for lenders to help householders improve the energy performance of their homes. Available from https://assets.publishing.service.gov.uk/government/uploads/system/uploads/attachment_data/file/936276/improving-home-energy-performance-through-lenders-consultation.pdf. Accessed 13 February 2022.

Bell, C. J., Nadel, S. and Hayes, S., 2011. On-bill financing for energy efficiency improvements. *A Review of Current Program Challenges, Opportunities and Best Practices*. Available from www.aceee.org/sites/default/files/publications/researchreports/e118.pdf. Accessed 13 February 2022.

Belloni, M., Giuzio, M., Kördel, S., Radulova, P., Salakhova, D. and Wicknig, F., 2020. The performance and resilience of green finance instruments: ESG funds and green bonds. Available from www.ecb.europa.eu/pub/financial-stability/fsr/focus/2020/html/ecb.fsrbox 202011_07~12b8ddd530.en.html. Accessed 13 February 2022.

Bertoldi, P., Economidou, M., Palermo, V., Boza-Kiss, B. and Todeschi, V., 2021. How to finance energy renovation of residential buildings: Review of current and emerging financing instruments in the EU. *Wiley Interdisciplinary Reviews: Energy and Environment*, 10(1), p.e384.

Boffo, R. and Patalano, R., 2020. *ESG Investing: Practices, Progress and Challenges*. OECD. Available from www.oecd.org/finance/ESG-Investing-Practices-Progress-Challenges.pdf. Accessed 1 October 2021.

Choi, W. Y. A., 2020. Can the green bond market reverse the tide of 'greenwashing'? An ecosystem mapping of international voluntary governance mechanisms (Doctoral dissertation), University of Geneva.

Cicmanova, J., Eisermann, M. and Maraquin, T., 2020. How to set up a one-stop-shop for integrated home energy renovation? A step-by-step guide for local authorities and other actors. Available from https://energy-cities.eu/wp-content/uploads/2020/07/INNOVATE_guide_FINAL.pdf. Accessed 13 February 2022.

Clayton, J., Devaney, S., Sayce, S. and van de Wetering, J., 2021. *Climate Risk and Commercial Property Values: A Review and Analysis of the Literature*. UNEP FI. Available from www.unepfi.org/wordpress/wp-content/uploads/2021/08/Climate-risk-and-real-estate-value_Aug2021.pdf. Accessed 9 February 2022.

Deschryver, P. and De Mariz, F., 2020. What future for the green bond market? How can policymakers, companies, and investors unlock the potential of the green bond market? *Journal of Risk and Financial Management*, 13(61), pp. 1–26.

EEFIG (Energy Efficiency Financial Institutions Group), 2015. Energy efficiency – the first fuel for the EU Economy. How to drive new finance for energy efficiency investments. Available from https://ec.europa.eu/energy/sites/default/files/documents/Final%20Report %20EEFIG%20v%209.1%2024022015%20clean%20FINAL%20sent.pdf. Accessed 13 February 2022.

European Banking Authority, 2021. Advice to the commission on KPIs and methodology for disclosure by credit institutions and investment firms under the NFRD on how and to what extent their activities qualify as environmentally sustainable according to the EU taxonomy regulation. Available from www.eba.europa.eu/sites/default/documents/files/document_library/About%20Us/Missions%20and%20tasks/Call%20for%20Advice/2021/CfA%20 on%20KPIs%20and%20methodology%20for%20disclosures%20under%20Article%20 8%20of%20the%20Taxonomy%20Regulation/963616/Report%20-%20Advice%20to%20 COM_Disclosure%20Article%208%20Taxonomy.pdf. Accessed 13 February 2022.

European Commission, 2019. Clean energy for all Europeans. Available from https://ec.europa.eu/energy/topics/energy-strategy/clean-energy-all-europeans_en. Accessed 13 February 2022.

European Investment Bank, 2020. Support for project development assistance. Available from https://ec.europa.eu/energy/sites/ener/files/documents/002_ralf_goldman.pdf. Accessed 13 February 2022.

Fuerst, F., Haddad, M. F. C. and Adan, H., 2020. Is there an economic case for energy-efficient dwellings in the UK private rental market? *Journal of Cleaner Production*, 245, p. 118642. https://doi.org/10.1016/j.jclepro.2019.118642.

Fuerst, F., McAllister, P., Nanda, A. and Wyatt, P., 2015. Does energy efficiency matter to homebuyers? An investigation of EPC ratings and transaction prices in England. *Energy Economics*, 48, pp. 145–156. https://doi-org.ezproxy.uwe.ac.uk/10.1016/j.eneco.2014.12.012.

Global Alliance for Buildings and Construction, 2020. Global status report for buildings and construction. Available from https://globalabc.org/sites/default/files/inline-files/2020%20Buildings%20GSR_FULL%20REPORT.pdf. Accessed 13 February 2022.

Green Finance Institute, 2021. Building renovation passports: Creating the pathway to zero carbon homes a report by the green finance institute's coalition for the energy efficiency of buildings. Available from www.greenfinanceinstitute.co.uk/wp-content/uploads/2021/03/GREEN-FINANCE-BUILDING-RENOVATION-final.pdf. Accessed 13 February 2022.

Hartenberger, U. and Lorenz, D., 2018. *Advancing Responsible Business in Land, Construction and Real Estate Use and Investment – Making the Sustainable Development Goals a Reality*. RICS, London. Available from www.rics.org/globalassets/rics-website/media/about/advancing-responsible-business-un-sustainable-development-rics.pdf.

Hartenberger, U., Toth, Z. and Sayce, S., 2019. Valuation checklist background explanation and guidance. Available from https://eemap.energyefficientmortgages.eu/wp-content/uploads/2018/11/Valuation-and-Energy-Efficiency-Checklist.pdf.

Högberg, L., 2013. The impact of energy performance on single-family home selling prices in Sweden. *Journal of European Real Estate Research*, 6(3), pp. 242–261. https://doi-org.ezproxy.uwe.ac.uk/10.1108/JERER-09-2012-0024.

Hyland, M., Lyons, R. C. and Lyons, S., 2013. The value of domestic building energy efficiency – evidence from Ireland, *Energy Economics*, 40, pp. 943–952. https://doi-org.ezproxy.uwe.ac.uk/10.1016/j.eneco.2013.07.020.

Kerr, N., Gouldson, A. and Barrett, J., 2017. The rationale for energy efficiency policy: Assessing the recognition of the multiple benefits of energy efficiency retrofit policy. *Energy Policy*, 106, pp. 212–221. https://doi-org.ezproxy.uwe.ac.uk/10.1016/j.enpol.2017.03.053.

Lashitew, A. A., 2021. Corporate uptake of the sustainable development goals: Mere greenwashing or an advent of institutional change? *Journal of International Business Policy*, 4(1), pp. 184–200. https://doi.org/10.1057/s42214-020-00092-4.

Lohse, R. and Zhivov, A., 2019. *Deep Energy Retrofit Guide for Public Buildings. Business and Financial Models*. Springer. https://doi.org/10.1007/978-3-030-14922-2.

McAllister, P. and Nase, I., 2019. The impact of minimum energy efficiency standards: Some evidence from the London office market. *Energy Policy*, 132, pp. 714–722. https://doi-org.ezproxy.uwe.ac.uk/10.1016/j.enpol.2019.05.060.

National Audit Office, 2016. Green deal and energy company obligation. Available from www.nao.org.uk/wp-content/uploads/2016/04/Green-Deal-and-Energy-Company-Obligation.pdf. Accessed 13 February 2022.

Panteli, C., Klumbytė, E., Apanavičienė, R. and Fokaides, P. A., 2020. An overview of the existing schemes and research trends in financing the energy upgrade of buildings in Europe. *Journal of Sustainable Architecture and Civil Engineering*, 27(2), pp. 53–62. https:// doi.org/10.5755/j01.sace.27.2.25465.

Purple Market Research (on behalf of BEIS), 2019. The impact of solid wall insulation on property value summary of findings of a pilot study conducted on social housing in Birmingham BEIS Research Paper Number 2020/029BEIS. Available from https:// assets.publishing.service.gov.uk/government/uploads/system/uploads/attachment_data/ file/913512/impact-solid-wall-insulation-property-value.pdf. Accessed 13 February 2022.

Rose, P., 2018. Certifying the 'climate' in climate bonds. *Capital Markets Law Journal*, 14(1), Ohio State Public Law Working Paper No. 458. http://dx.doi.org/10.2139/ssrn.3243867.

Rosenow, J. and Eyre, N., 2016. A post mortem of the green deal: Austerity, energy efficiency, and failure in British energy policy. *Energy Research & Social Science*, 21, pp. 141–144. https://doi.org/10.1016/j.erss.2016.07.005.

Rosenow, J. and Galvin, R., 2013. Evaluating the evaluations: Evidence from energy efficiency programmes in Germany and the UK. *Energy and Buildings*, 62, pp. 450–458. https://doi.org/10.1016/j.enbuild.2013.03.021.

Sayce, S. L. and Hossain, S. M., 2020. The initial impacts of Minimum Energy Efficiency Standards (MEES) in England. *Journal of Property Investment & Finance*, 38(5), pp. 435–447. https://doi-org.ezproxy.uwe.ac.uk/10.1108/JPIF-01-2020-0013.

Schröder, M., Ekins, P., Power, A., Zulauf, M. and Lowe, R., 2011. *The KFW Experience in the Reduction of Energy Use in and CO₂ Emissions from Buildings: Operation, Impacts and Lessons for the UK*. UCL Energy Institute, University College London and LSE Housing and Communities, London School of Economics, London. Available from https://sticerd. lse.ac.uk/dps/case/cp/kfwfullreport.pdf. Accessed 13 February 2022.

Sesana, M. M., Salvalai, G., Brutti, D., Mandin, C. and Wei, W., 2021. ALDREN: A methodological framework to support decision-making and investments in deep energy renovation of non-residential buildings. *Buildings*, 11(3), pp. 1–20. https://doi. org/10.3390/buildings11010003.

Toth, Z., Bertalot, L., Johnson, J., Leboullenger, D., Richardson, S. and Marijewycz, M., 2018. Roadmap for market implementation of the energy efficiency mortgage EeMAP. Available at https://eemap.energyefficientmortgages.eu/wp-content/uploads/EeMAP_ EMF-ECBC_D6.5_Final.pdf. Accessed 13 February 2022.

United Nations, 2015. The Paris agreement. Available from https://unfccc.int/sites/default/ files/english_paris_agreement.pdf. Accessed 13 February 2022.

Volt, J., Toth, Z., Glicker, J., De Groote, M., Borragán, G., De Regel, S., Dourlens-Quaranta, S. and Carbonari, G., 2020. *Definition of the Digital Building Logbook*. European Commission Publications Office. Available from https://data.europa.eu/doi/10.2826/480977. Accessed 13 February 2022.

Wilkinson, S. J. and Sayce, S., 2020. Decarbonising real estate: The evolving relationship between energy efficiency and housing in Europe. *Journal of European Real Estate Research*, 13(3), pp. 387–408. https://doi-org.ezproxy.uwe.ac.uk/10.1108/JERER-11-2019-0045.

Wrigley, K. and Crawford, R. H., 2017. Identifying policy solutions for improving the energy efficiency of rental properties. *Energy Policy*, 108, pp. 369–378. https://doi-org.ezproxy. uwe.ac.uk/10.1016/j.enpol.2017.06.009.

7 Low-tech versus high-technological solutions for a pandemic-adaptable society

Sara Wilkinson and Samantha Organ

7.1 Introduction

This chapter explores innovations in technology that could bring about significant change to the carbon footprint of existing buildings to address critical issues. Extreme times demand consideration of fresh new approaches. For example, many predict the end of fossil fuels as primary energy sources that facilitated the industrial revolutions of the 19th and 20th centuries in favour of biofuels, nuclear and renewable sources for the 21st century and beyond. Indeed, this is already happening through a combination of investment (public and private) and the impact of legislation and policy. In evidence, the UK's national electricity grid experienced 18 days of fossil-free energy generation in April 2020 (The Guardian, 2020). Can buildings derive sufficient energy from non-fossil sources, and will it need rapidly accelerated rates of retrofitting to achieve the change from fossil fuels and a reduced carbon footprint? Are there other benefits too – such as bioremediation of greywater in buildings? Other innovations in sensors, technology, artificial intelligence and robots may facilitate greater adoption of green infrastructure, which, if adopted 'en masse' in city centres will mitigate the urban heat island effect. One method of distinguishing these innovations is whether they are considered to be 'low tech' or 'high tech'. That is, whether they have little or no reliance on computerised technology (low tech); or whether features such as computer technologies, sensors or artificial intelligence (AI) are integrated into a building ('high tech') for performance optimisation. Drawing on innovations from a range of disciplines, this chapter sets out new technologies, new ideas and new ways of retrofitting existing buildings to deliver more sustainable and resilient outcomes.

Two key issues drive much innovation in the area of sustainable buildings:

a. Embodied carbon versus operational energy

Embodied carbon is defined by the Circular Ecology (2020a) as *'the carbon footprint of a material. It considers how many greenhouse gases (GHGs) are released throughout the supply chain and is often measured from cradle to (factory) gate, or cradle to site (of use)'*. Embodied carbon can also be measured in terms of the 'cradle to grave' framework. This framework includes material extraction, transport, refining,

DOI: 10.1201/9781003023975-9

processing, manufacture, the in-use phase (of the product) and, finally, its end-of-life profile. Industry and governments now acknowledge that embodied carbon emissions make up a significant portion of the emissions from the construction sector. Indeed, embodied carbon is estimated between 20% and 50% of the whole lifecycle (embodied + operational) carbon emissions of a new building (Circular Ecology, 2020a). This is a significant proportion and will increase as the thermal standards of new buildings improve, and operational carbon subsequently reduces. Embodied carbon is an important component of net zero carbon buildings (Circular Ecology, 2022) and construction projects and is included in the UK Green Building Council (UK-GBC) Net Zero Carbon Framework definition (UK-GBC, 2019) – for example, in recognition of its significance. Unless the embodied carbon footprint is included in the definition of net zero, it risks neglecting a large amount of upfront carbon emissions.

Taking a whole lifecycle carbon approach to guide our decisions in relation to building interventions is important. As we undertake works to our buildings, we inevitably add embodied carbon through the inclusion of new materials and systems. When specifying materials and systems, the lifespan and performance need to be considered, in addition to the impact these have on the wider building performance. But it is also increasingly important to compare the embodied carbon of the alternative materials for retrofits and other planned works, and through careful, considerate selection of materials and systems, it is possible to mitigate the amount of embodied carbon we add to existing buildings.

b. The three pillars of sustainability – social, environmental and economic

The second issue is the conceptual framework we use to frame our thinking about sustainability. The three pillars formed part of the 1987 UN Brundtland definition of sustainability in respect of inter- and intra-generation equity; that is, meeting the needs of the present generation without compromising the ability of future generations to meet their needs (Circular Ecology, 2020b). These three pillars are informally referred to as 'people, planet and profits'. When considering measures, stakeholders can evaluate the total costs and impacts through consideration of the three pillars and determine the overall best option. There are a number of sustainability decision-making tools in the built environment which adopt this framework (Khoshnava et al., 2018).

7.2 Principles for smart NZE/ZE retrofit to existing buildings

There has been some debate about what constitutes a NZE retrofit of existing buildings. The concept has roots in research from the 1970s (Esbensen & Korsgaard, 1977). Based on the definition outlined by the European Union's Energy Efficiency Directive (EED) (2018/844, 2012/27/EU and 2010/31/EU), it is the retrofitting of a building that leads to zero or very low energy demand. This energy demand is then primarily met through renewable energy sources, typically on-site or close by.

The principal desired policy outcome is to decarbonise the existing building stock and transform buildings into ones that are highly energy efficient through 'deep renovations'. This reflects indicators presented under Goal 7 of the UN's Sustainable Development Goals, particularly 7.3 which requires doubling the global rate of improvement in energy efficiency in 2030 (United Nations, no date (a)).

The EED requires EU member states to promote equal access to financing, consider affordability and address split-incentive dilemmas, and makes direct reference to the need to alleviate 'energy poverty' through renovation. Therefore, not only does the EDD relate to fuel poverty and energy poverty, a significant issue within the housing stock (Bouzarovski & Petrova, 2015), but it also consequently relates to the concept of energy justice (Gillard et al., 2017). In connection with this, the EED requires EU member states to promote equal access to financing, consider affordability and address split-incentive dilemmas. There is, therefore, a crucial role for NZE buildings to deliver on a number of key elements of EU policy.

An extension to NZE buildings is 'energy positive' buildings. For these buildings, energy surplus to occupant needs is generated through renewable sources and exported to the main energy network or to nearby connected buildings. To be considered 'energy positive', these buildings should produce surplus energy throughout the year rather than during specific periods such as in the summer (Magrini et al., 2020).

Given the size of the existing building stock, retrofitting to NZE is crucial to meeting carbon reduction and energy efficiency targets. There are principles and standards which can be adopted to facilitate the delivery of NZE buildings through retrofit. For example, the UK-GBC (2020) highlights 17 opportunities for net zero carbon framed using the RIBA Plan of Works Stages.

EnerPHit is a Passivhaus standard applied to existing buildings (Passivhaus Trust, no date), designed to take into account limitations and challenges that may be present in existing buildings. EnerPHit adopts the Passivehaus Planning Package (PHPP) design tool to facilitate informed design decisions (Passivhaus Institute, no date).

The EnerPHit standard requires that the building achieves a specified energy demand depending on the climate zone the building is located in. For example, for buildings in a cool-temperate zone such as counties in western Europe, this would be a maximum of 25 kWh/m^2 per annum for heating (Passivhaus Institute, 2016). This is a higher heating energy demand than the requirement for the Passivhaus standard for new buildings (i.e. 15 kWh/m^2 per annum), but significantly lower than the average space heating demand of the average home. For example, in the UK the average home has a space heating demand of approximately 145 kWh/m^2 per annum (Mitchell & Natarajan, 2020).

The low heating demand is achieved through high levels of insulation, high performance triple-glazed windows and doors, mechanical ventilation with heat recovery (MVHR) and airtightness of 1.0 air changes per hour. In contrast, for new houses, Passivhaus requires an airtightness of 0.6 air changes per hour, and the 2010 UK Building Regulations (Part L1A) require an airtightness

$10.0 \text{ m}^2/(\text{h.m}^2)$ at 50 Pa, although this is expected to reduce to $8.0 \text{ m}^2/(\text{h.m}^2)$ at 50 Pa for the 2021 regulations. At the time of writing, these proposed standards were released for consultation in January (closed in April 2021 (UK Gov, 2020a, b)

Beyond adopting standards such as EnerPHit, there are guiding principles for achieving net zero energy and carbon buildings. ARUP (no date) suggests a number of key considerations:

1. Passive building designs to produce less operational carbon and greater levels of occupant comfort;
2. Reusing materials and refurbishing buildings where possible, and use of prefabricated materials to minimise material waste. These approaches can limit embodied energy and carbon, and forms part of circular economy principles;
3. Switching from fossil fuels to electricity as the main power source, and ensuring the use of efficient services and systems with good controls to minimise operational energy and carbon;
4. Reducing operational energy and reducing peak loads through demand management. Supply emissions are high during peak energy periods and reduces infrastructure capacity. The use of smart building technology, automatic load shedding, renewable technology and/or battery technology can facilitate good demand management; and
5. Use on-site renewable technology where feasible and where this would deliver an appropriate solution in relation to lifecycle costs and carbon.

NZE are yet to become common practice. Currently, the perceived challenges of increasing the proportion of NZE buildings include the lack of awareness of NZE buildings; limited or no requirement for NZE buildings within building regulations and standards; actual or perceived higher up-front costs; the availability of affordable finance; and insufficient incentives (Ferrante, 2016).

There has been a burgeoning of technological innovation associated with lowering energy consumption and carbon emissions from buildings, contributing to NZE targets. Some of these innovations are considered to have the potential of being 'disruptive' in the sense of, if successful, creating a new market and disrupting current approaches and business models (Wilson et al., 2019). In the realms of architecture and the built environment, 'low-' and 'high-technology' have been defined in divergent ways. In this chapter, we take these terms to mean built environment innovations capable of delivering reductions in carbon emissions and energy consumption through integration into the building or built environment. This can be in the form of low-technology materials such as strawbales, or high-technology integrated systems such as algae building. However, the terms are not mutually exclusive and are likely to be subject to continual evolution. Ongoing reflection is crucial to ensure that any changes in technology are sufficiently understood and that they are utilised where they are relevant and appropriate to the context. This following sections outline some of the low- and high-technology innovations currently being adopted in the built environment.

7.3 Innovations in technology – bio tech

Some researchers and practitioners look to nature and so called 'bio technologies' to provide options for greater resilience and sustainability. This section highlights innovations and options in respect of on-site production of 'bio energy' in the form of algae biomass, followed by use of low energy construction materials: hempcrete, strawbale and rammed earth used in wall and floor construction. Each of these materials can be found in most countries and offer proven alternatives to energy intensive concrete.

7.3.1 Algae building tech

In 1839, Alexandre Becquerel discovered the photovoltaic (PV) effect; however, energy generated by PV was inefficient and prohibitively expensive until 1941, when Ohl invented the solar cell. Developments in battery storage, smart electricity grid management and greatly reduced costs, transitioned PV to a viable alternative to fossil fuels. In the 1950s, PV cost AS\$2,723.32 per watt in 2016 money. Slowly, then swiftly the cost of solar cells fell to less than AS\$1.14 per watt (The Guardian Sustainable Business, 2016). Sudden, disruptive and largely unpredictable technology shifts occur, making technologies viable and attractive (Davila et al., 2012). This occurred with solar and could happen for other 'new' renewable energy technologies. Indeed, an example of this was the introduction of the UK's Feed-in Tariff (FiT) in 2010. The FiT had a number of aims including changing public attitudes towards small-scale low-carbon electricity generation, such as the solar photovoltaic panels; accelerating technological development; and developing the supply chain and stimulating the market. The incentive saw a reduction in the cost of the technology, particularly photovoltaic panels (DECC, 2015). Global biomass energy production reached 88 GW in 2014 (Rosillo-Calle & Woods, 2012); as such bio-energy is no longer a transition energy source.

The BIQ House was constructed in Hamburg in 2013, in a cool temperature Northern European climate. Fifteen apartments are located over four floors plus a penthouse level, with 50–120 metres squared space and a gross floor area approximately 1,600 m^2 (Buildup, 2015). 200 m^2 of integrated photo-bioreactors (PBRs) in 120 panels, on two façades, generate algal biomass and heat as renewable energy resources in a low-energy residential building (see Plate 7.1). The façade panel system provides a thermally controlled microclimate around the building, noise abatement and dynamic shading (ARUP, 2016). Construction costs were approximately 5M Euros (Buildup, 2015).

How does it work? Microalgae are cultivated in flat panel glass 'bioreactors'. Each year, the bioreactor façade removes up to six tonnes of carbon dioxide (CO_2) by using flue gas from the gas burner to produce the biomass within the 'photo-bioreactors' (PBRs). Excess heat from these PBRs is used to pre-warm domestic hot water, warm interior spaces or can be stored for later use.

Sunlight, constant turbulence and the addition of CO_2 as a nutrient cause the microalgae to grow. This produces heat energy to heat the building, and biomass.

Plate 7.1 BIQ Building Hamburg
(Source: *Wilkinson et al., 2017*)

Up to 80% of this biomass is converted into methane via an external biogas plant. Methane is used to fuel the micro-combined heat and power unit to generate both heat and electricity. During the combustion process in the micro-CHP, additional heat and CO_2 are produced, and these are fed back into the system. The heat is recovered via a heat exchanger to contribute to the building's heating demands, and the CO_2 is used to supply the algae. Any surplus heat from the system can be stored in geothermal boreholes. The heat has 38% efficiency compared to 60%–65% for a conventional solar thermal source. The algae-derived biomass has 10% of efficiency compared to 12%–15% with a conventional PV (Buildup, 2015), so the performance currently does not match alternate renewable energy forms.

The associated heat production of about 40°C (150 kWh/m²/yr) is reintroduced to the system via a heat exchanger in the heating network or is stored in below ground geothermal boreholes. The boreholes store heat from 16 to 35°C depending on the season. When a higher temperature is required for heating and/or hot water, a heat pump propels the water back into the system. A unit is operated to provide the CO_2 nutrient (flue gas) required by the microalgae in the bioreactor façade and to cover the supply of hot water at 70°C or heating in the energy network (Buildup, 2015).

Microalgae absorb sunlight and the bioreactors also provide dynamic shading, with the amount of sunlight absorbed and shading, dependent on the algae density. When there is more sunlight, the algae grow faster providing more shading for the building (ARUP, 2016). According to ARUP (2016), the flat PBRs on the Hamburg building are highly efficient for algal growth and require minimal maintenance. The PBRs have four glass layers: a pair of double-glazing units creating a cavity, filled with argon gas to minimise heat loss.

Water temperature in the PBRs is controlled by the speed of the fluid flow through the panel, with lower flow rates allowing greater time for sunlight to warm the water as it passes through, and by the amount of heat extracted via heat exchangers in the central plant. The maximum temperature within the PBRs is around 40°C, as higher levels harm the microalgae. Temperature constraints pose challenges to applying the BIQ system in hot countries such as Australia. The relatively low maximum PBR temperature limits the practical use of the extracted heat to mainly a pre-heating function for other building systems. Further, the maximum growing temperature for the algae species used in Hamburg may limit panel use to cooler regions of hotter countries where air temperatures exceed 40°C. However, it is possible to use other algae species, which tolerate higher temperatures.

The total energy system conversion efficiency is 27% relative to the full available solar radiation incident on an unobstructed building roof (ARUP, 2016). PV systems yield an efficiency of 12%–15% and solar thermal systems 60%–65%, when placed optimally to capture the total available solar radiation. Total energy conversion of the BIQ algae system is lower than that of conventional solar hot water panels, the BIQ building's bio-responsive façade necessarily aims to provide energy directly to several building services systems, to provide additional energy benefits through summertime shading, and by providing a biomass stock for additional use.

Take up and acceptance of ABT requires an understanding and view of the system's benefits for owners, users and built environment professionals such as planners, building surveyors, project managers, contractors, quantity surveyors, certifiers property managers and facility managers (ARUP, 2016). However, solar energy innovation has occurred over 181 years to date, costs fell sharply from the 1950s to the 2010s and thus, innovation can transition viability dramatically in short timeframes.

7.3.2 Hempcrete

Another innovative material is hempcrete. Bio-based materials, derived from plant sources, have become increasingly popular, producing eco-friendly materials with a low carbon footprint. Bio-based materials made from renewable vegetable granulates allow materials to have a net carbon storage due to CO_2 fixation during plant growth (Colinart et al., 2012). Among these materials, hemp concrete, or hempcrete, is becoming more and more popular in construction because of its manufacture from renewable resources (plants), and its non-degradable characteristics over time (Amziane & Arnaud, 2013; Castel et al., 2016). In comparison with conventional building materials, other advantages of hempcrete are its lower density, excellent acoustic properties (Kinnane et al., 2016;), excellent moisture buffer capacity allowing for the control of interior environmental quality (Khan et al.,

2014) and excellent thermal insulation properties (Bennai et al., 2018). Hemp-crete is also considered to be 'carbon negative' through carbon sequestration due to the biogenic element of the hemp shivs (i.e. the carbon stored during the growth of the plant) and the non-biogenic element of the lime binder, which absorbs atmospheric carbon for the process of carbonation (Jami et al., 2019).

Hempcrete (Plates 7.2 and 7.3) is a low embodied carbon construction mate-rial, here used in wall construction, as a result of carbon storage during hemp growth, and due to the low quantity of binder required compared to traditional concrete. Two types of binders are commonly used in Hempcrete: Ordinary Port-land Cement (OPC) or lime. Manufacture of OPC and lime is carbon intensive, involving decarbonisation of limestone.

The thickness of walls has good thermal insulation qualities and results in energy efficient performance. Retrofits involving the reconstruction of walls should consider hempcrete as an alternate material to brick or concrete. There is a need for greater awareness and understanding across the built environment regarding the performance and general properties of hemp as building materials. The material offers opportunities to reduce embodied carbon and enhance general sustainability.

Plate 7.2 Hempcrete Wall construction Marrickville, NSW, Australia
(Source: Wilkinson)

Plate 7.3 Hempcrete walls internally Marrickville house, NSW Australia
(Source: Wilkinson)

Beyond hempcrete, there are other applications of hemp and lime products being adopted in construction to improve energy efficiency. This includes materials such as hemp-lime plaster, internal lime insulation (Tŷ Mawr, no date (a)) and insulated limecrete floors (Tŷ Mawr, no date (b)). Indeed, hemp-lime plaster has been shown to achieve good thermal properties (e.g. Agliat et al., 2020). The use of such materials may, however, require some additional upskilling and understanding among specifiers and contractors in how to apply these unfamiliar materials (Organ, 2020), and the management of occupant expectations if materials take longer to 'set'.

7.3.3 Strawbale

Straw has been used as a building material for centuries for thatch roofing and it is mixed with earth in cob and wattle and daub wall construction. Strawbales were used for building in the 1800s by settlers in Nebraska, US, shortly after the invention of baling machines. Straw is derived from grasses and is a renewable building material since its primary energy input is solar and it can be grown and harvested. Straw, the springy tubular stalk of grasses, such as rice and wheat, which are high in tensile strength, is composed of cellulose, hemicellulose, lignin and silica. As it breaks down in soil, waste straw can be used as mulch. Furthermore, different grasses have somewhat different qualities; for example, rice straw contains a considerable amount of silica, which adds density and resistance to decomposition. A study from Chile compares wheat straw and corn husk bales (as well as EPS) and although there are similar thermal properties, the compressive strength of corn husks exceeds wheat and EPS (Rojas et al., 2019).

Strawbale walls are highly resistant to fire, vermin and decay. The structural loadbearing capacity of strawbales is good and, in the loadbearing strawbale method, walls up to three storeys high have been constructed. In Australia, most strawbale construction uses a frame of timber or steel for the building structure to comply with the Building Code of Australia (BCA). In the UK, popular construction forms include loadbearing strawbales or timber frame, such as ModCell, a prefabricated timber and straw structural panel system which has been adopted for a range of applications including housing, education buildings, offices, and retail units (ModCell, 2021).

Strawbales (See Plate 7.4) have very low thermal mass, being composed, by volume, mostly of air. However, the cement and earth renders typically used on strawbales result in finished walls having appreciable thermal mass in the masonry 'skins' either side of the insulated straw core. With earthen renders a render skin of up to 75 mm can be achieved, providing significant thermal mass.

With regard to insulation properties, strawbales perform to very high levels and are among the most cost-effective thermal insulation available. Centimetre for centimetre, straw has similar insulation value to fibreglass batts and a typical strawbale wall has an R-value (i.e. the measure of thermal resistance) greater than 10 m^2K/W. Furthermore, dollar for dollar, the insulation value of a strawbale wall exceeds conventional construction. It is essential that roofs and windows are well insulated to maintain the overall performance of strawbale construction.

Another attribute of strawbale construction is its excellent cost-effective sound insulation, which contributes to the liveability of this form of construction. Further, as fire cannot burn without oxygen, given strawbales are tightly packed and covered with render to produce dense walls with a nearly airless environment,

Plate 7.4 Laying strawbale

the fire resistance of compacted straw is very good. In Californian bushfires that destroyed conventional structures, strawbale homes survived.

In 2002, the Commonwealth Scientific and Industrial Research Organisation (CSIRO), an Australian Government agency responsible for scientific research, on behalf of Ausbale and the South Australian fire authority, undertook tests to produce a two-hour fire rating. The three types of standard sized rendered straw-bales – earth; lime and sand; and lime, sand and cement, were subjected to a simulated bushfire front with a maximum heat intensity of 29 kW per square metre – an accepted standard under AS 3959, Construction of Buildings in Bushfire Prone Areas. Further, a completed wall has excellent resistance to vermin and the typical termite protection measures required in the BCA are generally sufficient. However, contractors need to prevent infestation of mice during construction when the bales are relatively unprotected. Most strawbale construction is coated with plaster or render which is adequate to keep animals out, and if they do manage to get inside, densely packed straw makes it hard for them to navigate through the space. During construction, the use of traps and baits can be considered to ensure the finished structure is sound and vermin-free.

Provided the straw is protected and not allowed to get waterlogged, strawbale buildings may have a lifetime of 100 years or more (Seyfang, 2010). The most detrimental factor affecting strawbale wall durability is long term or repeated exposure to water. After two or three weeks, the fungi in bales produce enzymes that break down straw cellulose if the moisture content is above 20% by weight. The best way to prevent rot in a finished structure is to create a waterproof, breathable wall. The survival of historic strawbale structures in Nebraska demonstrates their durability in climates with variable moisture and temperature.

Straw is a waste product which cannot be used for feed, unlike hay, and much is burned at the end of the season. Using straw for building reduces air pollution and stores carbon. The straw left over from building can be used as much so that, overall, there is minimal waste from using the material (see 'Waste minimisation'). To be sustainable in the long term, straw needs to be grown in ways that maintain soil quality. Strawbale walls usually require concrete footings which adds to the embodied energy of their construction. Rice straw is a by-product of irrigation agriculture that changes the flow and water balance of catchments in Australia's major river systems. Wheat straw is less water intensive.

Greenhouse gas emissions associated with strawbales are very low. For example, a tonne of concrete requires more than 50 times the amount of energy in its manufacture than straw. Using straw for building will store carbon that would otherwise be released but the amount sequestered per dwelling is relatively small. Straw's primary value is as an insulating material that enables houses to use less energy in operation and have lower carbon dioxide emissions over the building's life.

7.3.4 Rammed earth

Globally, there has been a range of earth-based construction adopted historically such as cob, clay lump and adobe. Variations of types, mixes and techniques were seen not just between countries but also regionally and locally. Rammed earth walls are formed by ramming selected aggregates: sand, silt, gravel and clay into place between flat panels known as formwork. Traditionally, construction workers repeatedly rammed the end of a wooden pole into the earth mixture to compress it; hence the name 'rammed earth'. Nowadays the pole has been superseded with a mechanical ram. Even so, it is perceived as a low-tech option, adopting traditional materials and methods of construction, working with the site and the climate.

Globally, thousands of unstabilised rammed earth buildings have survived many centuries of use. However, most contemporary rammed earth walls are built with cement as a stabiliser and are typically 300 mm thick for external walls and 300 mm or 200 mm for internal walls. Stabilised rammed earth includes around 5%–10% cement to increase the strength and durability of the material. Stabilised rammed earth walls are usually covered with an air-permeable sealer to also increase the durability of the material.

Most embodied energy or rammed earth is derived from quarrying the raw material and transportation to site. Therefore using on-site materials will reduce construction-related embodied energy. Rammed earth has excellent thermal mass but limited insulation, so depending on the building's location, insulation may be needed.

Wall colour is related to the earth and aggregate used. Ramming layer by layer can lead to horizontal stratification to the walls, which can enhance the overall appearance. It can be adopted as a feature or eliminated depending on the preference. Also, the aggregates can be exposed and special effects created by adding different coloured material in layers, and feature stones or objects, alcoves or relief mouldings. Unusual finishes can be formed by including shapes in the formwork that can be released after the wall has been rammed.

7.4 Innovations in low and high technology – AI and smart tech

We need our buildings to be good energy citizens. This means that they need to consume less energy through energy efficiency and energy optimisation. It is essential that we reduce their energy demands on national energy networks and contribute to those networks through on-site renewable energy generation. This can be achieved not only through digitalisation and the interoperation of key technologies but also through carefully modelling solutions on an individual building level and a larger building stock level. However, technologies should be selected where they represent value to the owner or occupiers to ensure usefulness and also to avoid adding unnecessary complexity and embodied carbon to a building.

Through industry and research collaborations, supported by funding, there has been a burgeoning of innovative technological approaches to enable the retrofitting of the existing building stock to deliver nZEBs (e.g. Elagiry et al., 2020). From projects such as RenoZEB which utilises prefabricated 'plug and play' systems to Energiesprong which uses a prefabricated package of measures financed by energy bill savings to deliver net zero energy buildings.

On a microscale, technologies such as 'smart thermostats' have been growing in popularity, although the realised energy savings resulting from these types of technologies remain highly variable, often dependent on user assimilation, settings and behaviours (Stopps & Touchie, 2021). However, when compared to conventional thermostats, smart thermostats may facilitate greater opportunities for energy and carbon savings through improvements to the user interface, enabling improved scheduling (Stopps & Touchie, 2021). Used in the right way, such technologies have the potential to empower building users to reduce energy consumption.

7.4.1 Energiesprong

Energiesprong (translated as 'energy leap') is a Dutch energy transition programme currently active in the Netherlands, Canada, France, Germany, Italy and the UK, as well as cities such as New York. The premise of the programme is to provide whole house refurbishments at scale, with the intention of building critical mass, stimulating the wider market to enable a good business case, innovative solutions and financing mechanisms.

Through 3D laser scanning of an existing house, a building information model is then developed, and this is used to produce technical drawings. These then guide the decisions and production of customisable prefabricated packages at 'flexible factories'.[1] On site, these complete refurbishment packages take between 1 and 10 days to install, after which the house should be net zero energy. The package of measures can incorporate prefabricated facades, insulated roofs with solar panels and smart heating, ventilation and cooling technologies. A long-term warranty of up to 40 years is provided against the indoor climate and energy performance of the property following the refurbishment.

Monitoring equipment is installed with a real-time display for occupant feedback about their energy consumption, and to enable any issues to be identified where expected energy consumption is exceeded.

The refurbishment of each home is financed through the savings on the energy bills against a payback period of 40 years. Through this, instead of paying energy bills the occupant pays an 'energy plan'.

7.4.2 Smart buildings

Buildings are increasingly embedding sensors and similar components to enhance their energy systems. This can result in the creation of more complex, networked cyber-physical systems which artificial intelligence can be designed to optimise.

Smart buildings incorporate multiple sensors, subsystems and actuators to facilitate automated monitoring and control of energy and the internal environment. In NZE buildings, it can help reduce energy consumption, thereby reducing the associated energy costs and carbon emissions. Algorithms are used to analyse datasets, and in the context of NZE buildings, AI will monitor, collect information, control, evaluate and manage energy consumption to enable energy savings whilst producing more comfortable internal environments for occupants. When used in conjunction with other technology such as 'big data', the cloud and the Internet of Things, AI can enable the active management of electricity grids beyond individual buildings through improving the accessibility of renewable energy systems in buildings (Farzaneh et al., 2021).

The particular challenges associated with smart buildings include concerns around security and data privacy. Smart buildings need to prove their long-term reliability and demonstrate their ability to contribute to what Farzaneh et al. (2021) call the 'betterment of society'.

7.4.3 Smart batteries

Low carbon, 'renewable' technologies are reliant on the availability of the energy source, such as sunlight and wind. By their nature, many of these energy sources are intermittent and may not coincide with when energy demand is greatest. For example, pre-Covid-19 in the UK, demand for electricity has typically been found to be greatest at around 8 a.m. and 5 p.m. (Department for Energy and Climate Change, 2014), varying slightly by season and between countries. Electricity consumption patterns altered slightly during Covid-19 lockdowns, resulting in the morning electricity demand peak shifting slightly later (National Grid, 2020). However, sunlight energy is greatest at noon, and we typically receive more power between 11 a.m. and 3 p.m. (NASA Earth Observatory, 2009).

For the renewable technologies integrated into our buildings, where a surplus of electricity is generated above demand, this can either be transferred to the national electricity grid or stored locally if the system includes battery storage. Where the peak times for renewable electricity generation does not complement peak consumption times, batteries offer the advantage of increasing the amount of renewable-generated electricity available.

Batteries add to the initial cost of a renewable technology system, but also potentially to the lifetime costs and embodied carbon/energy of the system, given that the current lifespan of such batteries are between 5 and 15 years. In contrast, the estimated lifespan of solar (photovoltaic) panels is currently around 30 years, necessitating the replacement of the batteries during the lifetime of the panels. However, a further consideration is the environmental impact of producing and disposing of batteries, which require the mining of minerals and metals. This results in environmental harm. Batteries need carefully disposing of and whilst recycling is an alternative, this can currently degrade their power.

There are various types of batteries, and the technology has rapidly developed over recent years (Wang et al., 2018). One development has been the integration of AI. AI can monitor patterns and vast amounts of data, then learning from this for energy optimisation (Boretti, 2021). Whilst the technology is being investigated for application on a city-scale such as NEOM, a new city being planned in Saudi Arabia, AI-enabled solar batteries are currently being used for individual buildings. In housing, such systems identify the recent energy generation from the solar (photovoltaic) panel and household's consumption patterns. It then takes into consideration the local weather forecast and the property's energy tariff. Using this information, it predicts the amount of solar energy the system is likely to generate and how much the household is likely to use.

Where the renewable technology does not produce sufficient electricity to meeting demand, coupling with an energy supplier who generates 100% of their electricity from renewable sources can help keep the operational carbon footprint of the building low. In housing, where occupants adapt to use electricity from their supplier at off-peak times and stored electricity from their battery at other times, this can also facilitate a reduction in peak loads and therefore strain on the national electricity grid, but also reducing utility bills where lower, off-peak tariffs are utilised (Octopus Energy, 2018). This has potential implications for 'energy justice' (Villavicencio Calzadilla & Mauger, 2018), which relates to the UN Sustainable Development Goals, particularly Goal 7 – ensure access to affordable, reliable, sustainable and modern energy for all (United Nations, no date (b)).

7.5 Delivering more sustainable resilient outcomes with new tech

There are many debates about whether new tech can deliver more or less sustainability and resilience in buildings. The argument often posited for new tech is that adoption of sensors, AI and computer technology can optimise building performance. This section highlights a new innovation that could facilitate more sustainable and resilient retrofit outcomes; here in respect of maintenance and upkeep of green walls; traditionally found to be a technology with high ongoing maintenance costs and needs.

7.5.1 Wallbot and Smart Green Walls

The benefits of urban green infrastructure (GI) are widely accepted and include urban heat island attenuation, increased biodiversity, reduced carbon emission, biophilia effects, provision of spaces for social interaction, attenuation of rainwater flooding and improved air quality (Wilkinson & Dixon, 2016). The opportunity for robotic technology to increase the uptake of green walls and facades whilst reducing OHS and maintenance costs is clear. A 2019 University of Technology Sydney (UTS) research project examined the advantages of using a new technology to inspect, monitor and maintain green walls on the side of infrastructure and buildings (See Plate 7.5 Wallbot). With climate change and increasing temperatures a stark reality for all of us, resilience and liveability, as well as sustainability, are greatly enhanced through the adoption and retrofit of GI. Despite the advantages of GI, adoption rates are low, mainly due to the perceptions of onerous, ongoing maintenance costs and, on high rise buildings, OH&S issues. Horizontal farming bots are well established, and from this the wallbot was conceptualised. An extensive literature review of existing robots and wall climbing mechanisms, power sources, pruning technologies and green waste collection, as well as sensor technology and costs, was undertaken prior to workshops with NSW industry stakeholders, Junglefy and Transport for NSW, who critiqued the proposed wallbot designs. Based on the experts' review, a prototype design based on a four-cable climbing mechanism was designed and prototyped at UTS. This wallbot has sensors to detect plant growth, health, shape, temperature and to create a 3D map of the plants that

Plate 7.5   Wallbot trails UTS, Sydney, Australia
(Source: Wilkinson et al., 2021)

can be updated over weeks, months and years. This research is now evolving to Smart Green Wall technology, where the control system for wallbot is integrated in the Building Management System (BMS) for maintenance and record keeping purposes. This technology will lower OHS risks and reduce ongoing maintenance needs and costs. It creates a new local industry for bot manufacture, bot installation and maintenance.

7.6 Conclusions

Adaptations, or building retrofits, can take many forms. This chapter has explored low and high tech options that are emerging as innovations in technology and their place in decision-making. The case of embodied carbon versus operational energy consumption was discussed with reference to taking a whole lifecycle, or 'cradle to grave', approach. Embodied carbon is important and forms part of net and nearly zero carbon targets for buildings and should be considered in decision-making. However, the three pillars of sustainability: social, environmental and economic factors also need to be taken into account. NZE principles and ways of achieving this in retrofits were also outlined and how the aim is to decarbonise existing stock through deep renovations. Going one step further, it is possible to make buildings carbon positive; where surplus energy is exported to the grid or energy network. Various options, schemes and frameworks to achieve NZE or carbon positive retrofits were outlined, with options being adoption of low, high and bio tech. Examples of low tech approaches are strawbale construction, whereas bio tech options include algae building technology with biomass grown in façade panels for conversion to biofuel. Other bio tech option includes hempcrete, where hemp is a binder which reduces the embodied carbon in the finished material. Approaches can use both low and high tech. An example of this is Energiesprong, which utilises technology to 3D scan buildings to inform the selection and adoption of prefabricated measures, enabling rapid NZE retrofits when on-site, funded by savings on energy bills. Technology is included to locally generate energy as well as for monitoring building performance. The latter can highlight potential issues where an energy performance gap arises.

Smart buildings use multiple sensors, subsystems and actuators to facilitate automated monitoring and control of energy and the internal environment. As such they optimise operational energy use and comfort. At precinct scale, AI can facilitate active management of electricity grids and improve accessibility to renewable energy systems. As a new innovation, there is less evidence of their performance over time, but this will change. AI is being incorporated into technology such as batteries, resulting in 'smart batteries'. These smart batteries analyse data on recent energy consumption patterns and weather forecasts to assess the amount of energy – for example, from solar photovoltaics is likely to be generated. This can not only support reducing peak loads to the electricity network, but it can also facilitate occupant behavioural changes, conscious carbon footprint reductions and lowering of energy bills where off-peak electricity tariffs are used to meet any deficit in on-site electricity supply.

Another innovation is the green wallbot. Green walls offer enhanced thermal performance and reduced energy consumption. However, a barrier to uptake includes perceptions of high maintenance costs and OH&S issue for maintenance staff. These are overcome through wallbot.

The well-being of future generations must be factored into the decisions we make now. Our current mainstream and previous methods of construction did not factor carbon; embodied and operational, into decision-making. This is changing. However, given that only 1%–2% is typically added to the total stock of buildings, it is in retrofit that our greatest chance to decarbonise our built environment lies. Finally as we emerge from Covid-19, with new knowledge and experience our ability to deliver more resilient building retrofits also offers the opportunity to rebuild economies.

Note

1 Defined as the ability of a factory to produce customisable packages (i.e. different dimensions and solutions) to reflect the heterogeneous housing stock.

References

Agliat, R., Marino, A., Mollo, L., Pariso, P. et al., 2020. Historic building energy audit and retrofit simulation with hemp-lime plaster – a case study. *Sustainability*, *12*(11), pp. 2–15. http://dx.doi.org/10.3390/su12114620.

Amziane, S. and Arnaud, L., 2013. *Bio-Aggregate-Based Building Materials: Applications to Hemp Concretes*. ISTE, London.

ARUP, 2016. World's first microalgae façade goes 'live'. Available from www.arup.com/News/2013_04_April/25_April_World_first_microalgae_facade_goes_live.aspx. Accessed 4 February 2016.

ARUP, no date. The road to zero. Net zero carbon buildings: Three steps to take now. Available from www.arup.com/perspectives/publications/research/section/net zero-carbon-buildings-three-steps-to-take-now. Accessed 5 July 2021.

Bennai F., Issaadi, N., Abahri K., Belarbi, A. and Tahakourt, A., 2018. Experimental characterization of thermal and hygric properties of hemp concrete with consideration of the material age evolution. *Heat Mass Transfer*, *54*, pp. 1189–1197. http://dx.doi.org/10.100 7%2Fs00231-017-2221-2.

Boretti, A., 2021. Integration of solar thermal and photovoltaic, wind, and battery energy storage through AI in NEOM city, *Energy and AI*, *3*(21), pp. 1–8, https://doi.org/10.1016/j.egyai.2020.100038.

Bouzarovski, S. and Petrova, S., November 2015. A global perspective on domestic energy deprivation: Overcoming the energy poverty – fuel poverty binary. *Energy Research & Social Science*, *10*, pp. 31–40. https://doi.org/10.1016/j.erss.2015.06.007.

Buildup, 2015. The BIQ House: First algae-powered building in the world. Available from www.buildup.eu/en/practices/cases/biq-house-first-algae-powered-building-world. Accessed 21 September 2016.

Castel, A., Foster, S. J., Ng, T., Sanjayan, J. G. and Gilbert, R. I., 2016. Creep and drying shrinkage of a blended slag and low calcium fly ash geopolymer Concrete. *Materials and Structures*, *49*(5), pp. 1619–1628.

Circular Ecology, 2020a. Embodied carbon assessment. Available from https://circularecology.com/embodied-carbon.html. Accessed 1 January 2021.

Circular Ecology, 2020b. What is sustainability? The three pillars of sustainability. Available from https://circularecology.com/sustainability-and-sustainable-development.html. Accessed 1 January 2021.

Circular Ecology, 2022. Net zero carbon buildings. Available from https://circularecology.com/net-zero-carbon-buildings.html. Accessed 15 February 2022.

Colinart, T., Glouannec, P. and Chauvelon, P., 2012. Influence of the setting process and the formulation on the drying of hemp concrete. *Construction and Building Materials 30*, pp. 372–380. https://doi.org/10.1016/j.conbuildmat.2011.12.030.

Davila, T., Epstein, M. and Shelton, R., 2012. *Making Innovation Work: How to Manage it, Measure it, and Profit from it*. FT Press, New York.

Department for Energy and Climate Change, 2014. Special feature – Seasonal variations in electricity demand. Available from https://assets.publishing.service.gov.uk/government/

uploads/system/uploads/attachment_data/file/295225/Seasonal_variations_in_electricity_demand.pdf. Accessed 1 July 2021.

Department for Energy and Climate Change, 2015. Performance and impact of the feed-in tariff scheme: Review of evidence. Available from https://assets.publishing.service.gov.uk/government/uploads/system/uploads/attachment_data/file/456181/FIT_Evidence_Review.pdf. Accessed 1 July 2021.

Directive (EU) 2018/844 of the European parliament and of the council of 30 May 2018. Available from https://eur-lex.europa.eu/legal-content/EN/TXT/PDF/?uri=CELEX:320 18L0844&from=EN. Accessed 2 July 2021.

Elagiry, M., Dugue, A., Costa, A. and Decorme, R., 2020. Digitalization tools for energy-efficient renovations. *Proceedings 2020*, 65(1). https://doi.org/10.3390/proceedings 2020065009.

Esbensen, T. V. and Korsgaard, V., 1977. Dimensioning of the solar heating system in the zero energy house in Denmark. *Solar Energy*, 19, pp. 195–199. https://doi.org/10.1016/0038-092X(77)90058-5.

Farzaneh, H., Malehmirchegini, L., Bejan, A., Afolabi, T., Mulumba, A. and Daka, P. P., 2021. Artificial intelligence evolution in smart buildings for energy efficiency. *Applied Sciences*, 11, p. 763. https://doi.org/10.3390/app11020763.

Ferrante, A., 2016. *Towards Nearly Zero Energy: Urban Settings in the Mediterranean Climate.* Butterworth-Heinemann, Oxford.

Gillard, R., Snell, C. and Bevan, M., 2017. Advancing an energy justice perspective of fuel poverty: Household vulnerability and domestic retrofit policy in the United Kingdom, *Energy Research & Social Science*, 29, pp. 53–61, https://doi.org/10.1016/j.erss.2017.05.012.

The Guardian, 2020. Britain breaks record for coal-free power generation. Available from www.theguardian.com/business/2020/apr/28/britain-breaks-record-for-coal-free-power-generation. Accessed 1 January 2021.

The Guardian Sustainable Business, 2016. Solar power what is holding back growth in clean energy? Available from www.theguardian.com/sustainable-business/2016/jan/31/solar-power-what-is-holding-back-growth-clean- energy?CMP=new_1194&CMP. Accessed 2 February 2016.

Jami, T., Karade, S. R. and Singh, L. P., 2019. A review of the properties of hemp concrete for green building applications, *Journal of Cleaner Production*, 239, pp. 1–17. https://doi.org/10.1016/j.jclepro.2019.117852.

Khan, B. A., Warner, P. and Wang, H., 2014. Antibacterial properties of hemp and other natural fibre plants: A review, *BioResources*, 9(2), pp. 3642–3659.

Khoshnava, S. M., Rostami, R., Valipour, A., Ismail, M. and Rahmat, A. R., 2018. Rank of green building material criteria based on the three pillars of sustainability using the hybrid multi criteria decision making method. *Journal of Cleaner Production*, 173, pp. 82–99. http://dx.doi.org/10.1016/j.jclepro.2016.10.066.

Kinnane, O., Reilly, A. and Grimes, J., 2016. Acoustic absorption of hemp-lime construction. *Construction and Building Materials*, 122, pp. 674–682. https://doi.org/10.1016/j.conbuildmat.2016.06.106.

Magrini, A., Lentini, G., Cuman, S., Bodrato, A. and Marenco, L., 2020. From nearly zero energy buildings (NZEB) to positive energy buildings (PEB): The next challenge – The most recent European trends with some notes on the energy analysis of a forerunner PEB example. *Developments in the Built Environment*, 3, pp. 1–12. https://doi.org/10.1016/j.dibe.2020.100019.

Mitchell, R. and Natarajan, S., 2020. UK Passivhaus and the energy performance gap. *Energy and Buildings*, 224, pp. 2–14. https://doi.org/10.1016/j.enbuild.2020.110240.

ModCell, 2021. ModCell straw technology. Available from www.modcell.com. Accessed 10 July 2021.

NASA Earth Observatory, 2009. Incoming sunlight. Available from https://earthobservatory.nasa.gov/features/EnergyBalance/page2.php. Accessed 2 July 2021.

National Grid, 2020. 4 ways lockdown life affected UK electricity use. Available from www.nationalgrid.com/uk/stories/grid-at-work-stories/4-ways-lockdown-life-affected-uk-electricity-use. Accessed 2 July 2021.

Octopus Energy, 2018. Agile Octopus – A consumer-led shift to a low carbon future. https://octopus.energy/static/consumer/documents/agile-report.pdf. Accessed 2 July 2021.

Organ, S., 2020. The opportunities and challenges of improving the condition and sustainability of a historic building at an international tourist attraction in the UK. *International Journal of Building Pathology and Adaptation*, 38(2), pp. 329–355. https://doi.org/10.1108/IJBPA-09-2018-0076.

Passivhaus Trust, no date. Passivhaus retrofit. Available from www.passivhaustrust.org.uk/competitions_and_campaigns/passivhaus-retrofit/. Accessed 10 July 2021.

Passivhaus Institute, 2016. Criteria for the passive house, EnerPHit and PHI low energy building standard. Available from https://passiv.de/downloads/03_building_criteria_en.pdf. Accessed 10 July 2021.

Rojas, C., Cea, M., Iriarte, A., Valdés, G., Navia, R. and Cárdenas-R, J. P., 2019. Thermal insulation materials based on agricultural residual wheat straw and corn husk biomass, for application in sustainable buildings, *Sustainable Materials and Technologies*, 17, pp. 1–5. https://doi.org/10.1016/j.susmat.2019.e00102.

Rosillo-Calle, F. and Woods, J., 2012. *The Biomass Assessment Handbook*. Earthscan, London.

Seyfang, G., 2010. Community action for sustainable housing: Building a low-carbon future. *Energy Policy*, 38(12), pp. 7624–7633.

Stopps, H. and Touchie, M. F., 2021. Residential smart thermostat use: An exploration of thermostat programming, environmental attitudes, and the influence of smart controls on energy savings, *Energy and Buildings*, 238, pp. 1–16, https://doi.org/10.1016/j.enbuild.2021.110834.

Tŷ Mawr, no date (a). Internal wall insulation system. Available from www.lime.org.uk/applications/retrofit-insulation-systems-for-old-buildings/internal-wall-insulation-system.html. Accessed 10 July 2021.

Tŷ Mawr, no date (b). Insulated limecrete floor. Available from www.lime.org.uk/applications/retrofit-insulation-systems-for-old-buildings/sublimer-limecrete-floor-insulation-system.html. Accessed 10 July 2021.

United Nations, no date (a). Ensure access to affordable, reliable, sustainable and modern energy. Available from www.un.org/sustainabledevelopment/energy/. Accessed 10 July 2021.

United Nations, no date (b). The 17 goals. Available from https://sdgs.un.org/goals. Accessed 10 July 2021.

UK Gov, 2021a. The future buildings standard. www.gov.uk/government/consultations/the-future-buildings-standard. Accessed 19 July 2021.

UK Gov, 2021b. Approved document L – Conservation of fuel and power Volume 1: Dwellings – Consultation. Available from https://assets.publishing.service.gov.uk/government/uploads/system/uploads/attachment_data/file/956100/AD_L_1.pdf. Accessed 10 July 2021.

UK Green Building Council, 2019. Net zero carbon buildings: A framework definition. Available from https://ukgbc.s3.eu-west-2.amazonaws.com/wp-content/uploads/2019/04/05150856/Net-Zero-Carbon-Buildings-A-framework-definition.pdf. Accessed 15 February 2022.

UK Green Building Council, 2020. Unlocking the delivery of net zero carbon buildings. Available from www.ukgbc.org/wp-content/uploads/2020/11/Delivery-Guidance-Report.pdf. Accessed 10 July 2021.

Villavicencio Calzadilla, P. and Mauger, R., 2018. The UN's new sustainable development agenda and renewable energy: The challenge to reach SDG7 while achieving energy justice. *Journal of Energy and Natural Resources Law*, 36(2), pp. 233–254. https://doi.org/10.1080/02646811.2017.1377951.

Wang, Q., Liu, W., Yuan, X., Tang, H., Wang, M, Zuo, J., Song, Z. and Sun, J., 2018. Environmental impact analysis and process optimization of batteries based on life cycle assessment. *Journal of Cleaner Production*, 174, pp. 1262–1273. https://doi.org/10.1016/j.jclepro.2017.11.059.

Wilson, C., Pettifor, H., Cassar, E., Kerr, L. and Wilson, M., 2019. The potential contribution of disruptive low-carbon innovations to 1.5 °C climate mitigation. *Energy Efficiency*, 12, pp. 423–440. https://doi.org/10.1007/s12053-018-9679-8.

Wilkinson, S., Carmichael, M. and Khonasty, R., 2021. Towards smart green wall maintenance and Wallbot technology. *Property Management*. doi:10.1108/PM-09-2020-0062.

Wilkinson, S., Stoller, P., Ralph, P., Hamdorf, B., Catana, L. N. and Kuzava, G. S., 2017. Exploring the feasibility of algae building technology in NSW. *Procedia Engineering*, 180, pp. 1121–1130.

Wilkinson, S. J. and Dixon, T. eds., 2016. *Green Roof Retrofit: Building Urban Resilience*. John Wiley & Sons, Chichester.

8 Repurposing and adaptation

Hilde Remøy and Sara Wilkinson

8.1 Introduction

Social and technological change will always affect buildings and how we design and, importantly for this chapter, use them. A key current example is the advent of driverless vehicles and a sharing economy model which is predicted to see a decline in car ownership. Currently, most cars globally are in use less than one hour per day, so we build structures to park them for 95% of the day. As this decline occurs, large amounts of car parking space will become redundant. Another example of this is change in industrial manufacturing processes; the decline of shipbuilding and manufacturing has left building stock and communities in decline at best and abandoned at worst. The so-called rust belt in North America, including cities of Detroit and Milwaukee, provides another good example of this decline. Elsewhere the retail sector and high street are suffering from the advent of online shopping. So, what can be done to repurpose and adapt this stock? This is on the assumption that we do not wish to demolish it.

As we know, some changes are unpredictable and fast, known as acute shocks in resilience parlance. Others are slow and ongoing or chronic. In 2020, the globe experienced a health shock in the form of Covid-19. The virus quickly turned into a global pandemic, a health crisis which spread rapidly and was exacerbated by international air travel. As a result, global travel shut down to essential travel only, with people required to quarantine on arrival in many countries. Soon economic impacts were felt; people were told to stay home and work and not to socialise outside the home. Retail switch to online shopping accelerated to minimise exposure to the virus, and restaurants turned to home delivery models. Socialising at sporting, cultural, music and arts events ceased, and at the time of writing are very limited. This is still having an impact on our existing building stock, and the full outcome is yet to be realised; however, we can see from previous change what can happen and, in this way, explore what might happen in the future.

This chapter looks at some innovative ideas for repurposing and adapting redundant stock for new uses which meet revised needs and demands. For example, urban food production, shared affordable and alternative housing are some of the options explored.

DOI: 10.1201/9781003023975-10

8.2 Innovative ideas for repurposing redundant stock

This chapter explores innovative ideas that have been adopted in various cities and countries for different land use types. Office, retail, industrial and residential conversions are covered. However, repurposing is not limited to these land use types exclusively; for example, a trend in many western industrialised countries is the increasing vacancy of church buildings. In the Netherlands, around 25% of the churches are vacant or have been sold off for new use (van der Breggen & de Fijter, 2019). Several new forms of use have been introduced, but as church owners and boards do not feel comfortable with just any new type of use, societal uses are often sought or commercial functions found suitable by the boards, such as libraries, book shops, hotels and restaurants. Since the outbreak of Covid-19 though, many shops and restaurants are closed, and tourists stay away from the hotels.

In Amsterdam, a former prison has been partly demolished and will be partly converted into residential and other functions. The demolished parts will be reused as much as possible in the new development. Parts of the former building will accommodate urban farming and will serve as a test site for circular developments; that is development (or in this case partial redevelopment) that aligns with the circular economic principles as set out in, for example Mangialardo and Micelli (2017). This is an excellent example of innovative sustainable adaptive reuse potential and it is likely that the short to long-term economic changes that arise from the Covid-19 pandemic will lead to land uses previously not considered viable becoming adopted. Prison and mental health asylums have posed challenges for repurposing in many locations due to the previous uses, and adaptive reuse needs to be carefully planned.

Industrial buildings are often located in areas that are becoming attractive for new activities as they become outdated for their original use and as they become subsumed into the city as that city grows swallowing areas that used to be on the city fringes, or if not on the fringe, in low value areas. Worldwide, stunning examples include the Elbphilharmonie, a concert hall in and on top of a former warehouse in Hamburg, the Tate Gallery of modern art in London, realised in a former power station, or the Circular Quay buildings in Sydney, now accommodating shops and restaurants. Other examples are housing with ground floor shops and restaurants in former warehouses Felix and Leidseveem in Rotterdam and a market hall in a former energy central in Oslo.

8.2.1 Office buildings

Former studies show the potential for delivering environmental, social and economic sustainability to urban areas by office retrofits and adaptive reuse, by upgrading the environmental performance of existing office buildings through within use adaptation, and by introducing new functions through adaptive reuse (Barlow & Gann, 1993; Tiesdell et al., 1996; Coupland & Marsh, 1998; Heath, 2001; Beauregard, 2005; Langston et al., 2008; Wilkinson et al., 2009; Bullen & Love, 2010;

Koppels et al., 2011; Remøy & van der Voordt, 2014; Remøy & Street, 2018; Wilkinson & Remøy, 2018).

Conversion of offices into residential has become common over the past 20 years, unlike the 1990s when it was often disputed as a development strategy. Then the controversy was rooted in a wide variety of issues. First of all, proof of the financial feasibility of the conversion investment costs was lacking. Secondly, conversion developments were compared to demolition and new-build where existing buildings were replaced by much denser developments. A third financial topic relates more to investors' psychology, through which they demonstrated a lack of knowledge about anything outside the main office market. Other hurdles for this type of conversion to become mainstream related to urban planning, land use and zoning plan issues, because urban areas, city centres and CBDs had strict zoning plans and were in general lacking societal infrastructure to support conversion into housing (Beauregard, 2005; Remøy & van der Voordt, 2014; Wilkinson & Remøy, 2018). Environmental sustainability was less focused on in both research and practice at the end of the nineties, but came to the fore in early 2000s (Remøy & Van der Voordt, 2009; Wilkinson et al., 2009; Wilkinson & Remøy, 2011; Wilkinson et al., 2014).

Following the 2008 global financial crisis, conversion of offices into residential has and is taking place on a larger scale in many countries and cities throughout the world, in several cases stimulated by government policy and regulations. In England and Wales, the permitted development rights (PDR) were extended around 2010 to stimulate the property market, land use change and building retrofits, taking account of the housing shortage in the market (Remøy & Street, 2018; Clifford et al., 2019). In the Netherlands, changes were made to the buildings decree and to the environmental law decree as part of the crisis and recovery measures following the global financial crisis (Heurkens et al., 2018; Remøy & Street, 2018). In 2019, 12,500 housing units were realised in the Netherlands through conversion projects, equal to 13% of the total housing production (CBS, 2020).

In a replacement market, office locations with new, high-quality office buildings are developed and organisations move from older buildings to new buildings. Older real estate is left behind. As demand is caused by movements and office space demand is not increasing, this real estate remains vacant and becomes obsolete (Remøy, 2010). In the Netherlands, this caused a higher vacancy than that which existed before the global financial crisis, driven by the willingness to invest in office real estate by investors, developers and local governments. The global financial crisis made the vacancy problem obvious and initiated conversions throughout the market. Buildings which were converted were to either smaller office buildings in the city centres, solitary buildings outside the city centres on non-office locations or larger office buildings as part of office locations in urban areas. Since 2015, Dutch office vacancy has been decreasing, and especially central office locations have experienced hardly any vacancy at all, limiting the drive for new conversions.

However, new energy-efficiency policies are expected to come into play in 2023, requiring an energy label C for offices according to the NTA8800 method

(RVO, 2020); similar provisions exist and are being strengthened in the UK, but here the minimum standards apply only to let buildings – whatever the use, with some exceptions. Assessing the possibilities for use after 2023, conversion is identified as an option, next to sustainable within-use adaptation. Anticipation of the new policy has already driven some conversions throughout the Netherlands. It is hard to say to what extent Covid-19 will add to this picture, although we know that there will be changes. In March 2020, the Dutch real estate market came to a halt; in the first half year of 2020 the demand dropped by 40%. In the Covid-19 landscape, organisations postpone accommodation decisions and wait for the end of the crisis (Dynamis, 2020). In a global survey held by Cushman & Wakefield, a majority of respondents indicated that they will prefer to work partly, though not fully, from home (Cushman & Wakefield et al., 2020). If this happens, it could lead to the next surplus of office space and again drive adaptive reuse.

The first Dutch examples of office to residential conversions, which took place around 2000–2010, were often examples of student housing and affordable housing, and in a few cases conversion to luxury residential schemes when the building was located in a good location, and for instance if the building had historic qualities. From 2010 onwards, a more diverse residential supply has been created through conversion. Currently, office to residential conversions show the same variety as new-build. What is interesting to see is that the existing building type and location steer the choice for the type of apartments and target groups for buying or renting. For instance, monumental office buildings have proved to be attractive for luxury housing, such as in Park Hoog Oostduin in the Hague.

Park Hoog Oostduin is a former Shell office in The Hague, the Netherlands, which was converted into a modern apartment complex with 230 luxury apartments.[1] High-quality apartments were developed through an approach of designing within the limitations of the original building, reusing existing features as much as possible. The structural grid of 1.8 metres enabled adaptability of the layout until late in the redevelopment process. But after completion, the building still has a high degree of adaptability. Much attention was paid to sustainability in the project. Following circular economy principles, at least 80% of the building materials were reused. Components and material were reused as much as possible within the project, and other materials were extracted and used in other projects. For example, large quantities of doors, door closers, wall lamps, office furniture and floor coverings have been reused in other projects and the old cooling installations and kitchen facilities have also been reused elsewhere.

New types of conversions, still in the experimental phase, are circular conversions, applying the principles of the circular economy to buildings that are to be converted, making sure that a minimum of new material is used, and that any new material which is used can be reused later (MOR, 2019). Moreover, new ideas for circular conversions focus on the future adaptability of the building, aiming to make a next conversion less complex, with less need for new building material, less construction and demolition waste and lengthening the lifespan of the building (MOR, 2019). Circular conversions often combine residential with different uses that support the residential function, and which contributes to the circular economy, for instance, by reusing household waste, or by introducing circular urban farming. Where conversion projects 20 years ago claimed to be sustainable because they were reusing an existing building, new conversions go further, and aim at becoming carbon neutral or even positive.

8.2.2 *Retail*

The retail sector is diverse is terms of age and scale. For example, some small corner shops or retail outlets in towns and villages date back centuries. These shops and businesses are owned often by families with no other retail outlets; often alterations and adaptations have occurred to ensure the building meets the then current owner and user needs. At the other end of the scale, retail buildings can be enormous. For example, with 3.163 million square feet, or 293,852.32 square metres, in one retail location, the Shinsegae destination flagship store in Centum City, Busan, South Korea, earned the title 'world's largest' in March 2009 (The Balance, 2021). However, even this figure palls when compared to the Dubai Mail shopping area of over 12 million square feet, or 1,114,836.48 square metres (or 50 soccer pitches), which makes it the largest shopping mall in the world based on total area. With over 1,200 shops, an ice rink, a SEGA game centre, a five-star hotel, 22 cinema screens and 120 restaurants and cafes, it also houses the largest indoor aquarium in the world and is the gateway to the Burj Khalifa, the world's tallest building. The diversity of this type of retail property is considerable. Shops range from small to large department stores; from food retailers to niche outlets; from global companies to local businesses.

The retail sector has been experiencing significant shifts because of developments in customer behaviour, such as the switch to online shopping which has been enabled by innovations in technology. For example, in 2021, in the Australian retail sector, and elsewhere, there are 'Customer Fulfilment Centres'. These centres enable retailers to outsource warehousing and shipping, relieving the online part of the business of the necessary physical space to store all products. Sellers send their products to the fulfilment centre, where robots find and pack items ordered on online for despatch to the buyer. Covid-19 shook the sector, accelerating the adoption of online shopping in various sectors of the population who felt unable to shop safely or those restricted by lockdown conditions and increasing adoption of robotics. In June 2020, the US retail sector forecast: 'A third of America's malls are going to shut permanently by 2021, according to one former department store executive, as their demise is accelerated due the pandemic'; the same article forecasts that 50% of US department stores will also close permanently (CNBC, 2020). Similar predictions exist in other countries. Smaller retail businesses have suffered disproportionately, partly due to the lack of online alternatives. As a result, vacancy in small high street retail shops is extremely high. Some of these locations will struggle to return to retail use in the short, medium or long term. Similarly, some retail outlets in the shopping centres or shopping malls are also experiencing high vacancies as a result of Covid-19.

The question is: what are alternate uses for this stock? Sections 8.2.3.1 to 8.2.3.6 of this chapter uses some illustrative case studies to explore alternate uses for retail stock. It is noted that some areas in some countries, the UK for example, local retail is experiencing more demand as people work from home during Covid-19 (Deloitte, 2021) and therefore there is variation in vacancy, supply and demand in the retail stock.

8.2.3 Shopping malls

Old suburban shopping malls are frequently repurposed into mixed-use developments including multi-family living, hotels, gyms or yoga studies, co-working office spaces, retail and entertainment venues, and sometimes even civic amenities such as libraries or post offices. According to JLL's survey (JLL, 2019) of 90 American super regional and regional malls that are undergoing, or have undergone, a significant renovation since 2014, 30% of them reported these renovations to be non-retail related. Apartment living, hotels and office buildings ranked as the top mixed-use developments at 41%, 33% and 26%, respectively. A study by Friedman LLP (2019) found 75% of US Mall owners were considering a repurposing project in the foreseeable future. Covid-19 has accelerated this trend.

As non-traditional tenants are added into the space, finding the right combination of occupants with good financial credential and who will be able to increase revenue is important. To appeal to the 72% of Millennials who prefer to spend money on experiences rather than material things, mall owners and developers often gravitate towards experiential mixed-use additions, including restaurants, movie theatres and entertainment centres that will drive foot traffic (Stoler, 2019).

Shopping malls are typically well located, near to highways, transport nodes (railway or light rail, tram or tube stations). Property owner Brookfield was spending US$149 million to redevelop a San Francisco State Galleria centre with healthcare, whole foods and sporting goods stores. In addition, a cinema was to be relocated into the Mall.[2] Covid-19 has hit cinemas hard in many countries with social isolation and lockdowns restricting opening and capacity. At the time of writing, it is unclear how well the cinema sector will rebound.

Another global commercial real estate and investment management company, Colliers, also support greater inclusion of wellness in retail, and stated 75% of US shoppers are more likely to re-visit malls, 72% would spend more and 61% would use restaurants more (Colliers, 2019). To be successful, adaptive reuse projects need to research the demographics and socio-economic profile of the catchment.

Covid-19 has put health and well-being to the top of the agenda and some US real estate stakeholders are looking at shopping malls being repurposed with digital imaging and outpatient services for dialysis, mammography, colonoscopy, CAT scans and MRIs. Doctors, GPs, podiatrists, and dermatologists would also be sited in the wellness centres. Complimenting these uses would be providers of supportive care such as sport medicine practitioners, physical and physiotherapists, nutritionists and massage therapists. Mental health could also be addressed with psychiatry, psychology, social workers and group therapists occupying smaller vacated retail spaces.

A further extension is preventative care with sport, yoga, dance and barre studio facilities being on site. This would create a need for activewear and footwear retailers – along with health dining venues. There is also potential to repurpose some parts of the outside areas such as parking areas into walking paths and spaces for events.

8.2.3.1 Malls to education use

Another option is to convert the Mall to education use. The variety of spaces from large open areas in larger retail stores to small retail space suit the diverse space needs of education providers. The Highland Mall in Austin Texas was purchased by Austin Community College in 2012 to create a new campus.[3]

Both these adaptive reuse options have high levels of social sustainability. The environmental sustainability is partly in the retention and reuse of an existing building, and partly in the adoption of sustainable technologies to lower operational energy use and embodied carbon.

8.2.3.2 Malls to residential use

Other options are to repurpose malls for residential and senior living facilities (Bloomberg, 2020). Here, the provision of the additional healthcare and well-being, sporting and healthy eating amenities are very attractive. The provision of the residential component is usually a 'top up' development added to the existing retail building. For example in Denver, developers are converting part of the 41-year-old Alderwood Mall outside Seattle into housing with a 300-unit apartment complex with underground parking, and Bloomberg claims it may be a sign of what might be a national trend (Bloomberg, 2020). Some retail will be retained in the development, with retail tenants retaining 90,000 square feet of space. When the new Alderwood reopens by 2022, the focus will have shifted dramatically. One of the Mall's anchor department stores, Sears, closed in 2019 and the apartment complex will be the new 'anchor' for the development (Bloomberg, 2020). This is an example of a partial adaptive reuse, replacing outmoded redundant land use with a viable social and economic alternative.

8.2.3.3 Malls to co-working use

Co-working, where people come to a space to rent a desk, has been gaining popularity in many sectors, especially start-up companies. Co-working spaces have the amenities of an office without tying small companies into lengthy leases. Although Covid-19 has resulted in most office workers working mostly from home in 2020, many people miss the energy of an office and being around other people. Co-working spaces could fill that gap to a greater extent going forward.

8.2.3.4 Impact of online shopping

Another Covid-19 impact in 2020 has been the acceleration of online shopping and click and collect globally. This was driven for some by anxiety of possibly contracting the virus whilst shopping; for others it was the only way to shop during a lockdown. As a result, the numbers of retailers offering online options grew. The question is: will this trend continue and will it be permanent? Some argue that many shoppers, especially younger people, want in-person connection and experiences in retail, with evidence from 2018 that millennials and Gen Z shoppers returning to bricks and mortar retailers. It seems there will always be some demand for face-to-face retail experiences. The trick for owners is to create an additional experience that cannot be achieved through online shopping.

It is said 90% of retail development in the next decade will involve existing buildings. It is imperative that developers understand the macro trends affecting demand in the sector to be ahead of the curve to attract tenants. In adaptive reuse projects, developers need to appreciate the reasons previous tenants' businesses failed in order to appreciate the new value of the redeveloped space.

8.2.3.5 Retail to urban farms and food production

Some cities and settlements potentially have issues around food security – that is when the supply chain may be disrupted or broken, thus leaving urban populations short of some foods. Another sustainability consideration is carbon food miles associated with the distance the food needs to travel from the point of growth to consumption and the mode of travel. As a result, the same food, grown in different locations will have a different carbon footprint. Some consumers now actively seek locally grown food to minimise their carbon footprint in respect of food. Another increasingly popular option is to grow some food in the cities: urban food production. The methods of growing food in cities have transitioned from allotments to vertical food production using hydroponic systems. Food is grown on racks under LED lighting. In this way the area of food grown far exceeds the physical footprint of the building. For example, a building 10 metres wide and 10 metres deep has a floor area 100 metres squared. If four racks can be accommodated per floor level, there are up to 400 metres squared of productive space on each floor level. Furthermore, because the plants are grown under controlled conditions and LED lights, there is less likelihood of crop failure. In some retail buildings, floor heights are very generous, and greater numbers of racks can be accommodated (Sananbious, 2021).

8.2.3.6 High street retail

Much retail stock comprises small high street shops. These are often two or three stories high and built in terraced form. They often have residential accommodation above the ground floor shop space. The challenge for much of this retail stock is that roads have become much busier, leading to on-street parking being removed and making it harder for people to access these shops. The noise of traffic also makes walking along these roads very unpleasant, with attendant air quality issues. Further many shops which were owner-occupied with the proprietor living above their shop units – but as national brands moved into the high streets with 'live out' managers, so the flats, which often have no independent access to the street, began to stand empty. The result is a high proportion of this stock is experiencing long-term vacancy.

What are the options for alternative uses? With traffic noise, conversion to residential use is a challenge. A successful retrofit would need to focus on noise reduction, so double or triple glazing would be desirable. This would also reduce energy use and lower the carbon footprint of the building. Could it be a possibility to convert this stock into affordable housing, maybe with small-scale offices or workplaces on the ground floor? It is an interesting challenge and a worldwide phenomenon. Good solutions are still being sought to address this challenge. In many older European cities, councils have banned cars from some parts of city centres, in order to provide a healthier and attractive environment. Whilst such moves are not new,[4] they have done, and undoubtedly will play a major role in enabling high streets to retain attraction, creating an ambience and health environment which promotes longer 'dwell' time and accommodates a mix of retail and hospitality. Key, however, is to enable good access via park and ride or penetration by public transport.

8.2.4 Industrial

Upcoming trends, new demands and a changing environment, together with the move from an industrial-driven to a service-driven economy and beyond, are factors that have contributed to the vacancy of industrial buildings which have lost their original function, some of which are considered industrial heritage buildings. Vacant buildings tend to quickly deteriorate because they are not heated, seldom well-maintained and, in turn, result in further financial depreciation (van Dommelen & Pen, 2013). Besides the negative developments of the buildings, vacancy can develop into a societal problem due to illegal occupancy, increased crime rates and risk of vandalism, leading in turn to an adverse effect on the surrounding community (Kraut, 1999; Douglas, 2006; Remøy et al., 2009). The municipality of Amsterdam (n.d.) declared that '*Empty buildings form a "dead" place in a neighbourhood, are quickly impoverished, and sometimes develop to become illegal landfills or worse. In short, empty buildings cause undesirable situations for a neighbourhood*' (cited in Remøy et al., 2009). Furthermore, many industrial buildings contain materials considered deleterious to human health such as asbestos, or the activities undertaken in the building involved deleterious materials such as lead and toxic chemicals.

Adaptation of a building takes place when one or more players are aware of the (potential) qualities of a building and/or its environment (Remøy, 2014). MMnews (2018) reported that '*heritage buildings have a unique character and attractiveness, but the opportunity for use of these often heavily deteriorated buildings are not always revealed*'.

Throughout the world, many examples can be found of successful conversion of industrial complexes and transformation of large-scale industrial sites. The former harbour areas in London in the UK are good early examples, but also the conversion of former car manufacturing sites and buildings, such as the Lingotto factory in Turin,[5] converted into a conference centre, and the Bicocca area in Milan transformed into a new hub for the creative and knowledge industry (Sacco & Blessi, 2009). In the Netherlands, fine examples of conversion of industrial buildings into a variety of new functions include the former Philips factories in Eindhoven and Rotterdam port warehouses.

8.2.4.1 Factories in Eindhoven

Since 1892, Eindhoven was the hometown of electronics company Philips. Throughout a century, Philips attracted many investors and companies to the city and facilitated the development of Eindhoven as a major and technological hub (Eurocities, 2015). Strijp-S is one of the largest former Philips grounds in Eindhoven and is positioned at the heart of the Brainport network, which has the ambition to invest in knowledge infrastructure and to realise attractive business climates for companies and employees. The StrijpS area was developed in the early 20th century. At the peak, the area provided work for 15,000–20,000 employees. In the early 1990s, Philips lost ground to both traditional competitors and new competitors from Asia. As Philips was forced to restructure, many production units were closed or relocated to few strategic locations, including the production units at Strijp-S (Van Winden et al., 2013).

In 2002, Philips transferred management of the area to Park Strijp Beheer: a pubic private partnership between the municipality and a developer (Van Winden et al., 2013) and redevelopment of the area could start. One of the aims of the redevelopment strategy was to give the area a new identity as a 'creative city', actively steering the development of creative entrepreneurship and a creative economy (Goulden, 2015). Twenty years later, Strijp-S is accommodating more than 700 companies with about 7,000 employees, with a strong focus on design and other creative activities. Over 1,000 residents live at Strijp-S and the area attracts many visitors to different types of events.

8.2.4.2 Warehouses in Rotterdam

The Port of Rotterdam is the largest port of Europe. However, due to the increasing scale of shipping, and changing perception towards environmental requirements, port-related industries continue to move away from the city centre and towards the sea. As a result, older port areas near the city centre with all the iconic industrial buildings – the Stadshavens (city port) - have become obsolete (Stadshavens, 2017).

The Municipality of Rotterdam, together with the Port of Rotterdam and market players, aims to realise innovative living and working in the Stadshavens in order to strengthen the economic structure of Rotterdam and its port (Stadshavens, 2017). In 2015, the Rotterdam Innovation District was launched, to catalyst new, creative industry and business developments. As industry is phased out, redevelopment plans are made for the areas, with the iconic industrial buildings playing an important role in the new developments.

8.3 Benefits of conversion

The conservation and reuse of industrial heritage generates several benefits, which can vary from commercial value to tangible and intangible community benefits (including the sense of history, educational and research value, spiritual value). Keeping industrial heritage can also be considered a duty to future generations. Like physical capital, cultural capital can be considered subject to decay if neglected. The asset value of existing cultural capital can be enhanced by investment in its maintenance or improvement; new cultural capital can be created by new investment. Social cost-benefit analyses could be conducted to analyse not only the financial value but also the cultural value of reusing industrial buildings. They also provide the opportunity to learn, understand and interpret our histories.

8.3.1 Urban vibrancy

Florida (2003) drew attention to the creative industry worldwide taking over former industrial sites. He highlighted the importance of the creative class for urban development and even stated that the job opportunities within the creative professions are a driving force behind economic growth (CBS, 2011). In line with the theory developed by Jacobs (1961), and translating this theory to an area level, conversion of industrial buildings to new, creative industry workplaces and creative hubs was found to have a positive influence on the urban liveliness and vibrancy. Her theory was updated to the 21st century by adding the importance of events and 'third places' (Brethouwer, 2018). By adapting or adding functions, a flow of people is provided, which ensures the liveability of the area (Jacobs, 1961). This way, the transformed industrial building in itself generates positive externalities, as opposed to the base-case scenario in which the building is left vacant (Koppels et al., 2011). Brethouwer (2018) found that conversion of obsolete industrial buildings into creative hubs, including work places, third places and event space, contributes to increased vibrancy in urban areas. However, the success of the conversion in terms of influence on the vibrancy depends also on the context. For instance, buildings located in areas separated from the urban fabric by physical boundaries and low accessibility, or that were socio-economically detached from their neighbourhood, did not have much positive influence on its surroundings. However, as part of larger developments, conversion of iconic buildings was found to boost developments and attract a diverse group of people, from residents and employees to artists, entrepreneurs and tourists.

8.3.2 Financial added value

Following Brethouwer's study (2018), Persoon and Remøy (2020) found a significant effect of conversion of industrial heritage buildings not only on the value of the building itself but also on the property values of nearby houses. The study analysed four converted industrial heritage buildings and 135,000 home sale transactions. Confirming Brethouwer's (2018) results, this study found that the results depended on the physical and the socio-economic context. As much of the financially advantageous impact of conversions was found to be capitalised by private homeowners, investments in conversion of cultural heritage are certainly, but not only, a public interest. Local governments should therefore consider conversion of industrial heritage buildings more actively also as a financial instrument in urban redevelopment.

8.3.3 Other uses

Typically, other building types are converted to residential land use. For example, former hospitals, such as Prince Henry Bay in Sydney New South Wales (NSW) saw the conversion of an infectious diseases and quarantine hospital into a master planned community with a variety of housing types.

The Prince Henry Hospital site is a heritage-listed former teaching hospital. It was designed by the NSW Colonial and the NSW Government Architect and built from 1881 by the NSW Public Works Department. Also known as Prince Henry Hospital and The Coast Hospital, it is owned by Landcom, an agency of the Government of NSW and added to the NSW State Heritage Register in May 2003. The land has cultural significance for the indigenous Aboriginal going back at least 20,000 years with dated sheltered occupation sites. Further development occurred during the early colonial period and then in the 1880s, the Colonial Hospital was located here to deal with the health impacts of a cholera outbreak; the prevailing thoughts being the coastal location benefitted from breezes to remove germs and infections. In addition, the site was sufficiently far from Sydney to discourage patients from leaving the hospital and wandering into the city. Further expansion of the hospital occurred during the two world wars and the name changed to Prince Henry Hospital. A number of locally and nationally significant people have been associated with the buildings and site over its history. In 1988, the closure of the hospital was announced and 11 years later in 1999, a masterplan with private housing, aged care and selected medical facilities was announced. The redevelopment included restoration of heritage buildings on the site. The cultural and physical history of the site and buildings was well documented.

This site presented complex political and architectural challenges and became the subject of intense community concern. Landcom has a public benefit mandate, and a plan was developed to deliver innovative solutions to achieving density while maintaining amenity, beauty and social cohesion on a large and significant site. The masterplan set out a new residential and community precinct that balanced the old and the new, open space and built form, private and public uses, to create a showcase of sustainable coastal urban renewal. The cultural and community benefits are significant: 80% of the site is retained in public hands; improved access to Little Bay Beach; facilities for seven community groups and a 1500 m^2 community centre and a new Rescue Helicopter Service facility off site. Heritage issues were addressed, with the site listed on the State Heritage Register,

19 heritage buildings and landscape items conserved, and the historic Flowers Ward restored and adapted as a Nursing Museum to acknowledge the site's history of healing. Residential buildings were required to be energy rated using Australia's National Home Energy rating tool, NatHERS, and to achieve a minimum 4.5 NatHERS rating and around 90% of demolition material was recycled.

Warehouses have been subject to considerable adaptive reuse into residential development in many countries, especially those located in docks which became redundant when cargo was transported into larger container ships during the 20th century and docks with a deeper draft was required. Many of these warehouses were located close to city centres and had attraction waterfront areas which were converted in mixed use developments to support the new residential use. The scale of the Victorian warehouses suited conversion to four and five storey apartments typically. Many had heritage listing which meant the external appearance changed very little. The aesthetic became a popular industrial clique. The building size and configurations lent themselves to compliance with the construction and building regulations, and the location meant the costs of conversion were recovered in the apartments' property value.

Other conversions include a particularly unusual conversion and adaptive reuse of a Danish Jaegersborg water tower in 2004 into a mixed-use development with student housing provided in 36 apartments.[6] The water tower function is retained in a 2000 cubic metre tank located above the apartments. In 2015, Polish brick-built farm buildings in Leszno were converted into a residential building for the elderly with onsite healthcare facilities.[7] In both cases, the new uses are integrated into the existing buildings in ways which complement the original design and demonstrate the prevailing design thinking at the time of the conversion. The designers considered the occupants' physical and social needs in the provision of spaces for meeting and socialising in the developments. The end results are revitalising buildings and the surrounding areas with new uses, in these cases, affordable housing.

8.3.4 Historically important conversions

Two well-known examples are the amphitheatre in Lucca from the 2nd century and the canal-houses in Amsterdam from the 17th century. The amphitheatre has gone through several adaptations, and now, 1,900 years later, what is left of the original structure is merely its spatial configuration; the theatre scene is now a piazza, and the buildings around the piazza have taken the place of rows of seats. The Amsterdam canal-houses are quite young compared to the theatre in Lucca, and here next to the spatial configuration, also the image of the facades and heights of the buildings are kept, though these were also adapted several times. The functions of the buildings have changed a number of times together with the interior floors and the rear facade of the buildings. As such they are excellent examples of ongoing utility, ability to adapt to new uses and construction that is durable yet repairable: in short, resilient building retrofits.

8.4 Conclusion

Repurposing, adapting and converting buildings is not a new phenomenon, conversions have taken place everywhere and at all times, and on different scales contributing to today's beloved historical cities and buildings.

Societal, technological and economic developments lead to new use patterns and requirements, whether the changes happen slowly, for example as results of a changing climate, or through evolving ways of working and living, or as results of revolutions or crises, like the 2008 financial crisis or the current Covid-19 crisis. Recent changes in the industrial, retail and office markets are leading to changes in functions and changes in use requirements. Online shopping, already growing and boosted by the Covid-19 crisis, has led to vacancy in shopping malls, high streets and other retail real estate. Higher quality and sustainability requirements, new technology and flex-working, again ongoing developments since the 1990s but accelerated by the Covid-19 crisis, lead to new user preferences for offices.

Real estate which can be adapted to accommodate changing use requirements contributes to urban resilience (Wilkinson & Remøy, 2018). Typically, large-scale retail is repurposed to entertainment centres, sports and health facilities, high street retail is adapted to accommodate housing, workshops and neighbourhood amenities. Offices are frequently converted into housing (Remøy, 2010) because the location and building requirements are similar. Industrial buildings and areas are reused for different purposes, for living, working, leisure and retail. With the growth of the creative industry, creative hubs that combine flexible workplaces, cafes and restaurants, gyms, neighbourhood amenities and housing have proven to be valid catalysts for repurposing buildings and areas, activating also the existing surrounding environment. This kind of development, together with new concepts of the sharing, circular and donut economy (e.g. Ramani & Bloom, 2021) are upcoming concepts that could contribute even more to urban resilience, by adapting the use to the existing built environment – first we shape our buildings, thereafter they shape us.

Notes

1 See details at www.v2com-newswire.com/en/newsroom/categories/residential-architecture/ press-kits/3571-03/park-hoog-oostduin-the-netherlands
2 See www.brookfieldproperties.com/en/search.html?BP.com[query] =san%20francisco%20 Galleria
3 See what has been done at www.highereddive.com/news/in-a-former-shopping-mall-austin-community-college-sees-a-new-way-to-learn/580320/
4 Norwich City Centre in the UK, which was pedestrianised in 1967, claims to be the first move of its type in the world see www.norwich.gov.uk/info/20397/london_street_50
5 See www.atlasofplaces.com/architecture/lingotto-factory/
6 See https://urbannext.net/jacgersborg-water-tower/
7 See https://www10.aeccafe.com/blogs/arch-showcase/2017/10/04/elderly-healthcare-and-residential-building-with-hotel-and-restaurant-in-leszno-poland-by-na-no-wo-architects/

References

The Balance, 2021. What is the world's largest retail store? Available from www.thebalanc-esmb.com/largest-retail-stores-2892923. Accessed 15 February 2022.

Barlow, J. and Gann, D., 1993. *Offices Into Flats*. Joseph Rowntree Foundation, New York.

Beauregard, R. A., 2005. The textures of property markets: Downtown housing and office conversions in New York City. *Urban Studies, 42*(13), pp. 2431–2445.

Bloomberg, 2020. A case for turning empty malls into housing. Available from www.bloomberg.com/news/articles/2020-06-30/a-case-for-turning-empty-malls-into-housing. Accessed 15 February 2022.

Brethouwer, M., 2018. The added value of creative residencies. Faculty of architecture and the built environment (MSc thesis), TU Delft.

Bullen, P. A. and Love, P. E. D., 2010. The rhetoric of adaptive reuse or reality of demolition: Views from the field. *Cities, 27*(4), pp. 215–224. https://doi.org/10.1016/j.cities.2009.12.005.

CBS, 2011. *Onderzoeksrapportage Creatieve Industrie*. CBS, Den Haag/Heerlen.

CBS, 2020. 12,5 duizend woningen door transformatie van gebouwen in 2019. Available from www.cbs.nl/nl-nl/nieuws/2020/44/12-5-duizend-woningen-door-transformatie-van-gebouwen-in-2019. Accessed 2 February 2021.

Clifford, B., Ferm, J., Livingstone, N. and Canelas, P., 2019. Understanding office-to-residential permitted development. In B. Clifford, J. Ferm, N. Livingstone and P. Canelas (Eds.), *Understanding the Impacts of Deregulation in Planning: Turning Offices Into Homes?* (pp. 35–45). Palgrave Pivot, Cham.

CNBC, 2020. A third of America's malls will disappear by next year, says ex-department store exec. Available from www.cnbc.com/2020/06/10/a-third-of-americas-malls-will-disappear-by-next-year-jan-kniffen.html. Accessed 26 January 2021.

Colliers, 2019. Summer 2019: Retail spotlight report the fountain of wellness in retail. Available from www.colliers.com/en/research/2019-summer-us-retail-spotlight-report.

Coupland, A. and Marsh, C., 1998. *The Cutting Edge 1998; The Conversion of Redundant Office Space to Residential Use. RICS Research*. University of Westminster., London.

Cushman & Wakefield and Centre for Real Estate and Urban Analysis, 2020. *Workplace Ecosystems of the Future*. Cushman & Wakefield, Chicago, IL.

Deloitte, 2021. What next for the high street? Part 2. Available from https://www2.deloitte.com/content/dam/Deloitte/uk/Documents/consumer-business/deloitte-uk-what-next-for-the-high-street-part2.pdf. Accessed 15 February 2022.

Douglas, J., 2006. *Building Adaptation*. Butterworth-Heinemann, Oxford.

Dynamis, 2020. *Sprekende cijfers kantorenmarkten 2020*. Dynamis, Amersfoort.

Eurocities, 2015. Strijp S: Turning the relocation of a leading company into an urban rejuvenation success story. Available from www.cultureforcitiesandregions.eu/culture/resources/Case-study-Eindhoven-Strijp-S-WSWE-A3AGYV. Accessed 15 February 2022.

Florida, R., 2003. Cities and the creative class. *City & Community, 2*(1), pp. 3–19. https://doi.org/10.1111/1540-6040.00034.

Friedman LLP, 2019. Adaptive Reuse, the process of repurposing a shopping mall. Available from www.friedmanllp.com/insights/adaptive-reuse-the-process-of-repurposing-a-shopping-mall. Accessed 15 February 2022.

Goulden, L., 2015. *Smart Strijp-S Vision, Creative Innovation Works*.

Heath, T., 2001. Adaptive re-use of offices for residential use: The experiences of London and Toronto. *Cities, 18*(3), pp. 173–184. https://doi.org/10.1016/S0264-2751(01)00009-9.

Heurkens, E., Remøy, H. and Hobma, F., 2018. Planning policy instruments for resilient urban redevelopment: The case of office conversions in Rotterdam, the Netherlands. In S. J. Wilkinson & H. Remøy (Eds.), *Building Urban Resilience through Change of Use* (pp. 39–56). Wiley-Blackwell, Chichester.

Jacobs, J., 1961. *The Death and Life of Great American Cities*. Random House, New York.

JLL, 2019. The future of retail is . . .. Available from http://link.jll.com/Thefutureofretailis. Accessed 15 February 2022.

Koppels, P. W., Remøy, H. T. and El Messlaki, S., 2011. The negative externalities of structurally vacant offices: An exploration of externalities in the built environment using hedonic price analysis. 18th Annual European Real Estate Society Conference: ERES 2011, Eindhoven, The Netherlands, 15–18 June 2011. ERES, Eindhoven.

Kraut, D. T., 1999. Hanging out the no vacancy sign: Eliminating the blight of vacant buildings from urban areas. *New York University Law Review*, 74(4), pp. 1139–1177.

Langston, C., Wong, F. K., Hui, E. C. and Shen, L. Y., 2008. Strategic assessment of building adaptive reuse opportunities in Hong Kong. *Building and Environment*, 43(10), pp. 1709–1718. https://doi.org/10.1016/j.buildenv.2007.10.017.

Latitude, no date. 9 of the biggest shopping malls around the world. Available from www.28degreescard.com.au/travel-inspiration/9-of-the-biggest-shopping-malls-around-the-world.html. Accessed 15 February 2022.

Mangialardo, A. and Micelli, E., 2017. Rethinking the construction industry under the circular economy: principles and case studies. In *International Conference on Smart and Sustainable Planning for Cities and Regions* (pp. 333–344). Springer, Cham.

MMNews, 2018. Herbestemming Vraagt Om Ondernemers Met Lef. Available from www.herbestemming.nu/actueel/herbestemming-vraagt-om-ondernemers-met-lef. Accessed 15 February 2022.

MOR, 2019. Modular office renovation. Available from www.tudelft.nl/solarurban/research/projects/modular-office-renovation. Accessed 15 February 2022.

Persoon, T. and Remøy, H., May 2020. De toegevoegde waarde van herbestemming van industrieel erfgoed. *Real Estate Research Quarterly*, pp. 1–10. Available from www.vogon.nl/artikelen/vogon-publicaties/item/download/59_ad768cb2669dfbddb47d1b4693f-88cad. Accessed 15 February 2022.

Ramani, A. and Bloom, N., January 2021. The donut effect: How Covid-19 shapes real estate. *Stanford Institute for Economic Policy Research – Policy Brief*, pp. 1–8.

Remøy, H., 2010. *Out of Office, a Study of the Cause of Office Vacancy and Transformation as a Means to Cope and Prevent*. IOS, Amsterdam.

Remøy, H., 2014. Preserving cultural and heritage value. In S. J. Wilkinson, H. Remøy and C. Langston (Eds.), *Sustainable Building Adaptation: Innovations in Decision-making* (pp. 159–182). Wiley-Blackwell, Chichester.

Remøy, H., Schalekamp, M. and Hobma, F. A. M., 2009. Transformatie van kantoorterreinen; Een stappenplan. *Property NL Research Quarterly*, 8(4), pp. 22–28.

Remøy, H. and Street, E., 2018. The dynamics of "post-crisis" spatial planning: A comparative study of office conversion policies in England and the Netherlands. *Land Use Policy*, 77, pp. 811–820. https://doi.org/10.1016/j.landusepol.2016.12.005.

Remøy, H. and Van der Voordt, D. J. M., 2009. Sustainability by adaptable and functionally neutral buildings. SASBE. In Verhoeven, M. and Fremouw, M. Delft, Publicatiebureau Bouwkunde. SASBE'09 book of abstracts. 3rd CIB International Conference on Smart and Sustainable Built Environments, 15–19 June 2009, 150 blz, Delft, The Netherlands.

Remøy, H. and van der Voordt, T., 2014. Adaptive reuse of office buildings into housing: Opportunities and risks. *Building Research & Information*, 42(3), pp. 381–390. https://doi.org/10.1080/09613218.2014.865922.

RVO, 2020. Vernieuwd energielabel woningen en gebouwen. Available from www.rvo.nl/sites/default/files/2020/08/infographic-vernieuwd-energielabel-woningen-en-gebouwen-nta-8800.pdf. Accessed 15 February 2022.

Sacco, P. and Blessi, T. G., 2009. The social viability of culture-led urban transformation processes: Evidence from the Bicocca District, Milan. *Urban Studies*, 46(5–6), pp. 1115–1135.

Sananbious, 2021. Vertical farming. Available from www.sananbious.com/vertical-farming. Accessed 25 February 2021.

Stadhavens, 2017. Get involved in M4H. Available from www.rotterdammakersdistrict.com/. Accessed 15 February 2022.

Stoler, M., 5 November 2019. Adaptive reuse, the process of repurposing a shopping mall. Available from www.friedmanllp.com/insights/adaptive-reuse-the-process-of-repurposing-a-shopping-mall. Accessed 15 February 2022.

Tiesdell, S., Oc, T. and Heath, T., 1996. *Revitalizing Historic Urban Quarters*. Hartnolls Ltd, Watford.

van der Breggen, M. and de Fijter, N., 2019. *Niemand zit te wachten op een lege kerk. Maar herbestemming is ingewikkeld*. Trouw, Amsterdam.

van Dommelen, S. and Pen, C. J., 2013. *Cultureel Erfgoed Op Waarde Geschat*.

Van Winden, W., De Carvalho, L., Van Tuijl, E., Van Haaren, J. and Van Den Berg, L., 2013. *Creating Knowledge Locations in Cities: Innovation and Integration Challenges*. Routledge, London.

Wilkinson, S. J., James, K. and Reed, R., 2009. Using building adaptation to deliver sustainability in Australia. *Structural Survey*, 27(1), pp. 46–61.

Wilkinson, S. J. and Remøy, H., 2011. *Sustainability and Within Use Office Building Adaptations: A Comparison of Dutch and Australian Practices*. Pacific Rim Real Estate Society, Gold Coast, Bond University.

Wilkinson, S. J. and Remøy, H. (Eds.), 2018. *Building Urban Resilience Through Change of Use*. John Wiley & Sons, London.

Wilkinson, S. J., Remøy, H. and Langston, C., 2014. *Sustainable Building Adaptation, Innovations in Decision Making*. Wiley-Blackwell, London.

9 Heritage

Learning from and preserving the past

Sara Wilkinson and Shabnam Yazdani Mehr

9.1 Introduction

Heritage buildings are important to remind us of our history, and the need to conserve and preserve therefore remain important. There are many ways in which these buildings physically embody resilience and they can be retrofitted and adapted sustainably. Many heritage buildings adopted, what we now consider to be sustainable materials and technologies, as these buildings predate industrialised methods of production and the reliance of high levels of energy and mechanisation for operation. Therefore, there is much to be learned from them. Over time some buildings are adapted and retrofitted retaining their class of use, whereas others are converted, or undergo adaptive reuse. At this point issues around the place and location, or 'genius loci' and authenticity become important. These changes are a result of the legal, technological, social, economic and environmental drivers prevailing in the location at that point in time. This chapter explores what we can learn from the heritage stock to make their retrofit resilient; and what we can learn and transfer into the retrofit of other, non-heritage stock. A model for assessing adaptive reuse of heritage buildings and a checklist for identifying and preserving 'genius loci' in adaptive reuse are proposed.

9.2 What is a heritage building?

'Heritage building' does not have an unequivocal definition. It has several components, or attributes, which co-exist to varying degrees. These include:

- cultural and historic values,
- intrinsic and in use values,
- symbolic values and
- the relationship between the building and location (Wilkinson & Remoy, 2017).

A building may have little architectural heritage value as a property but have value because a momentous event transpired there. Heritage buildings fulfil important demands for cultural experiences and leisure and create benefits for

DOI: 10.1201/9781003023975-11

tourism (Willson & McIntosh, 2007). Heritage buildings also have other positive economic impacts, such as generating higher rental and occupancy levels (Jayantha & Yung, 2018).

Many terms – preservation, renovation, restoration and conservation – are used when talking about heritage. What do they mean? Furthermore, what conditions and responsibilities do they engender? Preservation aims to halt deterioration and ensure no further changes occur. It places requirements on materials and methods, as final appearance is no longer the most important factor. Rather, the aim is to retain as much of the existing building fabric as possible. Renovation means 'to make new again'. In buildings subjected to numerous interventions, the question is: *to which period do we renovate?* Outcomes depend on available documentation of original construction methods, materials and layouts and create a challenging debate among stakeholders (Wilkinson & Remoy, 2017). Heritage conservation is defined as ensuring the cultural significance of heritage items is maintained over time, and while changes to use may be necessary it is important to retain the heritage significance (NSW Heritage Office & Royal Australian Institute of Architects, New South Wales Chapter & Heritage Council of New South Wales, 2008. The heritage significance of a building must be investigated and fully evaluated to assess the extent of this significance. At this point a conversation management plan can be developed. Principles adopted in conservation are continuing to use the building, repairing rather than replacement, making reversible alterations and making a visual distinction between the old and new work. Avoiding precise imitation of architectural details, ensuring alterations are sympathetic and respect the ageing process, respecting previous alterations but discontinuing previous unsound practices where found, stabilising problem areas and respecting the building's context and location are other accepted principles to adhere to (NSW Heritage Office & Royal Australian Institute of Architects, New South Wales Chapter & Heritage Council of New South Wales, 2008).

9.2.1 Adaptive reuse

Adaptive reuse involves a wide range of activities, from maintaining a heritage building because of its specific features and values to changing the function of the building, either wholly or partially, for other uses (Douglas, 2006; Conejos et al., 2013; Plevoets, 2014; Wilkinson et al., 2014; Yazdani Mehr, 2018). Accordingly, although restoration, conservation, preservation and maintenance have different definitions, these activities can be part of the adaptive reuse practice.

According to the Nara document on authenticity: 'the diversity of cultures and heritage in our world is an irreplaceable source of spiritual and intellectual richness for all humankind' (ICOMOS, 1994). Based on the Nara document, heritage properties should be preserved for the present and future generations due to their important role in demonstrating the cultural identities of communities. Furthermore, many authors consider heritage buildings as finite resources which need adaptation in order to guarantee their continuous use for future generations (Tomaszewski, 2007; Burman, 2008; Jokilehto, 2008; Laenen, 2008; Lehne, 2008;

Schofield, 2008; Bold et al., 2017). Bold et al. (2017:6) expressed heritage as a tool which manages cultural diversity, develops participation and improves the quality of life. They further defined heritage as 'a catalyst for participation and social improvement'. This statement implies the social importance of heritage buildings for communities and the need for their adaptation.

Adaptation of heritage buildings requires an understanding of the identity and values assigned to them (ICOMOS, 1994). However, as heritage buildings may have a series of values for different people or groups (ICOMOS, 2013), there can be challenges and even conflicts in adaptation. The main aim of adaptation is to preserve the specific qualities of heritage buildings. Jokilehto (2008) believed that one of the most important reasons behind adaptation is the attached identity and values of heritage buildings. The Venice Charter (1964) stated that heritage buildings must be delivered to future generations in 'the full richness of their authenticity', implying little or no change to the former, original state. Authenticity is considered a complex phenomenon (Del Río Carrasco, 2008), which needs to be defined theoretically and valued practically. Authenticity is defined as 'the essential qualifying factor concerning values' (ICOMOS, 1994), indicating that authenticity compliance is critical in addressing all tangible and intangible values. Bold et al. (2017) noted one of the difficulties in the adaptation of heritage buildings is related to authenticity.

9.2.2 Authenticity

Authenticity of spirit is a concept considered by stakeholders. ICOMOS (2004:21) stated, 'in the evaluation of a monument not only the oft-evoked historic fabric but also additional factors ranging from authentic form to authentic spirit play a role'. The statement acknowledges the concept of 'genius loci' which, by implication, includes spirit and sense of place. Holden (2012) defined these two terms by expressing spirit of place as outside of us, whilst sense of place is inside of us and can be provoked by a landscape. 'Genius loci' gives identity to a place and thus distinguishes different places (Kepczynska-Walczak & Walczak, 2013) because it is related to events or actions, tangible or intangible values. For example, a prison might not be architecturally important; however, if it housed an important prisoner, such as Nelson Mandela, it becomes historically significant, and thus possesses genius loci. As such, it is both genius loci and authenticity that make a place unique. These statements imply the connectedness of authenticity and genius loci, as well as the importance of these concepts in the assessment and adaptation of heritage buildings. However, dilemmas can arise in practice in regard to these concepts in the adaptation of heritage buildings. According to Kepczynska-Walczak and Walczak (2013), genius loci has been overlooked in scientific analysis and is a difficult concept to define due to its abstract character and internal complexity. Bold et al. (2017) believed that authenticity is also an evolving concept and its meaning is becoming diluted. It is what Gallie (1955) calls an essentially contested concept; 'authenticity' means all things to all people. These notions indicate that although genius loci and authenticity are important in the adaptation of heritage buildings, their effect or perceived importance may be reducing in practice, due to a lack of knowledge and understanding. In acknowledgement of the complex nature of genius loci and authenticity, this chapter provides a holistic view related to these terms in order to ascertain how genius loci and authenticity can be preserved in the adaptation of heritage buildings.

9.2.3 Preservation

Preservation can be defined and perceived in different ways throughout the world. In the UK, preservation is described as works aimed at impeding or retarding deterioration, which is more commonly used in the context of structures. However, in the US, preservation is synonymous with conservation work (Douglas, 2006). The term 'preservation' was defined by Paul Philippot in 1972 as '*being equivalent . . . to conservation or restoration – can be considered, from this point of view, as expressing the modern way of maintaining living contact with cultural works of the past*' (Wong, 2016:11). As such, preservation comprises a wide range of activities to keep a heritage building in its place, respecting its existing features and integrity.

Preservation is generally defined as maintaining the structure of a heritage building in its current state by slowing down its deterioration (Department of Planning Sydney, 1995; ICOMOS, 2013) through the application of suitable repair methods. Under such circumstances, the intention is for preservation to extend the life-cycle of such heritage buildings. Douglas (2006) defined preservation as '*the act or process of applying measures necessary to sustain the existing form, integrity and materials of a historic property*'. Douglas (2006) believed that the preservation of heritage buildings focuses on the maintenance and repair of existing historic materials and structure as they have evolved over time. It includes protection and stabilisation measures. This complements Morris and Webb's 1877 Manifesto for the Society for the Protection of Ancient Buildings, which advises to 'stave off decay by daily care, to prop a perilous wall or mend a leaky roof' (Slocombe, no date; SPAB, 2017). Therefore, the preservation of a heritage building is usually carried out either to prolong the life of the heritage building or to make it functional.

Preservation tries to avoid the high level of natural decay (Brooker & Stone, 2004; Scott, 2007). Accordingly, preservation aims at retaining the condition of a heritage building which appears to have historical importance, even though some parts of the building may be damaged. Aplin (2002) believed that, although preservation involves comprehensive works and programmes to both maintain the fabric and mitigate damage, it is considered similar to maintenance.

Preservation of heritage buildings is considered the best and most accurate form of preserving buildings in situ, which may include protection, renovation and replacement of deteriorated materials (Ch'ng, 2010). From this perspective, preservation could support the notion of adaptive reuse.

Ruggles and Silverman (2009) state the aim of the preservation of heritage buildings is to retain and respect those parts of the past that still have meaning for the present, and possibly future generations, ensuring the ongoing use of heritage buildings. Regarding the role of heritage values in the preservation of heritage buildings, Riegl, the general conservator of the Central Commission of Austria during the first decade of the 20th century, defined several values associated with heritage buildings (Yazdani Mehr, 2019a). Riegl believed that for the preservation of a heritage building, it is essential that the values of the building's period must be defined (Yazdani Mehr, 2019a). It was further stated by Penna (2014), the preservation of a heritage building depends on its associated values that are defined by a specific group of people from a specific place and time in terms of what they wish to preserve and convey to future generations. Therefore, preservation of heritage buildings can be considered as a subset of adaptive reuse, aiming at retaining the current form, material and structure of heritage buildings and respecting their associated heritage values.

9.2.4 Conservation

Conservation means all the processes of looking after a place so as to retain its cultural significance, which includes maintenance and may, according to circumstances, include preservation, restoration, reconstruction and adaptation, or commonly be a combination of more than one of these (Boito & Birignani, 2009; ICOMOS, 2013). Aplin (2002) believed that conservation is a way of caring for the natural and cultural significance of a place to keep that significance for the present and future generations. Conservation reflects on the need to retain the inherited built environment, enabling future generations to appreciate, learn about and value their history (Shinbira, 2012). Therefore, conservation comprises a wide range of activities that aim to protect a heritage building, retaining its heritage values and extend its physical life. Shinbira (2012) believed conservation is a broad concept that runs from least intervention to greatest; which is, from maintenance to alteration of a heritage building. Based on this statement, conservation can be considered as another subset of adaptive reuse. Yung et al. (2014) further noted the adaptive reuse of heritage buildings is increasingly known as a sustainable approach to conservation.

Although conservation comprises a wide range of activities, it is defined by Azizi et al. (2016) as a way of retaining an existing building without altering or destroying its character, even though repairs or changes might be necessary. They believed that conservation work in the form of proposing changes to a heritage building should be distinguishable from the original existing form and structure, in line with the SPAB's 1877 Manifesto and current guidance (Slocombe, no date; SPAB, 2017). Based on this expression, the conservation of heritage buildings could be considered as the preservation of as much original fabric as possible, enhancing and respecting their existing character and features. Accordingly, conservation can be regarded as a way of prolonging the life of heritage buildings.

Similarly, as with preservation, conservation begins with identifying and assessing the cultural significance of a place before making any decisions related to the proposed changes (Kwanda, 2015). However, on the contrary to preservation, which aims at retaining a heritage building and its values in situ, Shinbira (2012) believed that conservation may involve the renovation of old structures to fulfil their original function through contemporary standards or even adapting them to new uses. Under this definition, a heritage building could be stripped down to its historic façade to act as frontispiece for a new function. Kwanda (2015) considers conservation as intervention that uncovers a previously hidden object or reveals the true nature of a heritage asset, such as removing a darkened varnish from a wooden statue, revealing the authentic appearance of the original state. These examples show the broad context of conservation.

Douglas (2006) considered the following seven main reasons behind the conservation of heritage buildings:

1. Cultural: conserving a heritage building due to its cultural significance/values.
2. Educational: conserving a heritage building as a learning resource.
3. Tourism: conserving a heritage building to attract visitors/tourists.
4. Historic variety: conserving a heritage building to retain an urban area's character.
5. Economic: conserving a heritage building to create new jobs.
6. Legal: conserving a heritage building to comply with local and national planning policies and legislation.
7. Technical: conserving a heritage building to minimise potential repairs in future.

Therefore, conservation comprises a degree of beneficial change, ensuring the survival of heritage buildings and their associated values.

9.3 Influential factors in adaptive reuse

In order to determine whether adaptive reuse is applicable to all existing buildings, authors have categorised various factors in environmental, social, economic, legal, political, locational, functional and technical categories, all of which effect decision-making in relation to undertaking adaptive reuse (Latham, 2000; Douglas, 2006; Langston, 2008; Bullen & Love, 2011a, 2011b; Wilkinson, 2011; Conejos et al., 2013; Wilkinson et al., 2014; Wilkinson & Remøy, 2018). The authors have grouped these categories in different ways. As an example, Wilkinson (2011) considered these categories under building adaptation theory, and then in 2014 (Wilkinson et al., 2014), these categories were considered in terms of drivers and barriers for adaptive reuse. In 2018 (Wilkinson & Remøy, 2018), these categories were grouped as opportunities and risks of adaptation. Acknowledging the importance of these categories in adaptive reuse, in this chapter, these are grouped in terms of drivers and challenges. 'Driver' is defined in the Oxford Dictionary (2018) as '*a factor which causes a particular phenomenon to happen or develop*'. As such, drivers promote and develop further adaptive reuse of existing buildings and often respond to a building's obsolescence, and thus postpone obsolescence in buildings. The term 'challenge' is selected because these factors and categories do not hinder the adaptation in terms of being a barrier, whilst the challenges need to be addressed properly to prevent a building from being impacted by obsolescence.

9.3.1 Drivers of adaptive reuse of heritage buildings

A range of drivers has been identified in the literature as promoting the adaptive reuse of heritage buildings. One, or a combination of drivers, may uphold the adaptive reuse of buildings, although the existence of numerous drivers will strengthen the motivation for undertaking adaptive reuse. There may be an overlap between these categories of drivers. For example, an existing building near public transport is desirable for adaptation in terms of factors contained in the multiple categories of locational, environmental, social and economic. As another example, the factor of governmental financial assistance can be considered a driver in both economic and political categories. Table 9.1 shows categories of drivers of adaptive reuse of heritage buildings identified in the reviewed literature and research undertaken by Yazdani Mehr (2019b).

As identified from the literature review, most authors focused on social factors, whilst none specifically considered technical or physical drivers of adaptation of heritage buildings. Whilst the number of factors do not reflect the actual importance of categories, it may indicate stakeholders' perceived importance of those drivers that have driven the adaptive reuse of heritage buildings over time.

9.3.2 Challenges to adaptive reuse of heritage buildings

Different factors identified in the literature pose a challenge to the adaptive reuse of heritage buildings. One, or a combination of challenges, may hinder the adaptive reuse of buildings to some extent. There may be an overlap between these challenges. For example, attaining the desired standard in an existing building can be an environmental and technical challenge to adaptation. Table 9.2 demonstrates the challenges to adaptive reuse of heritage buildings, based on research by Yazdani Mehr (2019b).

For the adaptation of heritage buildings, in the literature the economic, legal and technical challenges featured most prominently, whilst the least focus is on the locational, physical and environmental challenges. Although economic issues are one of the most challenging categories, Ball (2002) stated that the high costs of adaptation can be ignored when measured against the environmental and social advantages of adaptation. Ball (2002) contends that the high costs of adaptation can be disregarded when measured against the environmental and social advantages of adaptation. This is the case with adaptive reuse of heritage buildings, which preservation of social values in the form of adaptive reuse is more important than considering the cost of adaptation. So, even though economic factors may be the biggest challenge in the planning and designing phases of adaptation, this issue needs to be debated to preserve heritage buildings and their cultural values. However, high costs of adaptation, without financial incentives or funds, may lead to the building being left vacant for a long period of time, especially for privately owned buildings.

Table 9.1 Identified drivers of adaptive reuse

Drivers	Factors	Total factors
Environmental	– Reduction of greenhouse gas emissions (Langdon, 2008; Armitage & Irons, 2013; Wilkinson et al., 2014; Organ et al., 2020) – Saving the embodied energy of existing buildings (Binder, 2003; Clark, 2008; Historic England, 2020) – Reduced energy use (Bullen & Love, 2011; Armitage & Irons, 2013; Historic England, 2020) – Reduced resources consumption (Bullen & Love, 2011; Wilkinson et al., 2014; Historic England, 2020) – Reduced waste Historic England, 2020)	5
Social	– Being on a heritage list (Ball, 2002; Irons & Armitage, 2011; Wilkinson et al., 2014) – Preservation of heritage values and significance (Bullen, 2007; Bullen & Love, 2011a, 2011b; Irons & Armitage, 2011; ICOMOS, 2013) – Demand for various functions (Bullen & Love, 2011) – Community demand and request for adaptation (Bullen & Love, 2011) – Preserving the skills and human efforts of the original builders (Bullen & Love, 2009; Koslow, 2010) – Reduced the negative visual impact of existing buildings (Yau et al., 2008) – The collective sense of ownership – Community sense of ownership and support (Conejos et al., 2013)	8
Economic	– Economic advantages of adaptive reuse in terms of tourism and providing employment (Aplin, 2002; Brown, 2004; Clark, 2008) – Reduced costs of new construction (Clark, 2008; Wilkinson et al., 2014) – Financial incentives (Aplin, 2002; Clark, 2008; Koslow, 2010; Bullen & Love, 2011)	3
Legal	– Being on a heritage list (Ball, 2002; Irons & Armitage, 2011; Wilkinson et al., 2014) – The collective sense of ownership – Compliance with contemporary building standards and regulations	3
Political	– Amalgamation – Government support – Government policies (Aplin, 2002; Clark, 2008)	3
Physical Locational	No physical driver was identified for the adaptive reuse of heritage buildings in literature review – Providing market opportunity due to the location (Bullen & Love, 2011) – Matching with streetscape aesthetically and visually (Bullen & Love, 2011) – The position of an existing building on the site (Bullen & Love, 2011)	`3
Technical	N/A – Technical drivers of the adaptive reuse of heritage buildings were not specifically identified in the literature review. However, there is an overlap between technical and environmental drivers.	`

Table 9.2 Identified Challenges to adaptive reuse

Drivers	Factors	Total factors
Environmental	– Attaining the desired levels of standards (O'Donnell, 2004; Bullen & Love, 2011)	2
Social	– Adaptation based on green standards (Koslow, 2010) – The high number of stakeholders (Douglas, 2006; Wilkinson et al., 2014; Hussein, 2017) – Awareness about adaptive reuse among the community (Bullen & Love, 2011)	3
Economic	– Being on a heritage list (Ball, 2002; Irons & Armitage, 2011; Wilkinson et al., 2014) – Risks and uncertainties in adaptive reuse projects (Reyers & Mansfield, 2001; Shipley et al., 2006; Koslow, 2010) – Lack of financial support – Lack of accurate estimation of the required budget for adaptive reuse (Bullen, 2004; Bullen & Love, 2011) – The lower possibility of securing loans for reused buildings (Koslow, 2010) – Incorrect timing of incentives (Koslow, 2010) – The high cost of adaptation (Australian Government Productivity Commission 2005; Shipley et al., 2006; Bullen, 2007; Wilkinson et al., 2014) – Required adaptation work and proposed use – The decline in public sector budget (Australian Government Productivity Commission, 2005) – Finding new elements through an adaptive reuse project	9
Legal	– Receiving approvals for any work on heritage listed buildings (Douglas, 2006; Bullen & Love, 2011) – Planning restrictions and/or building regulations (Bullen & Love, 2011; Wilkinson et al., 2014; Wilkinson & Remøy, 2018) – Compliance with building codes and regulations (Bullen & Love, 2011) – Compliance with heritage guidelines (Bullen & Love, 2011) – Being on a heritage list (Ball, 2002; Irons & Armitage, 2011; Wilkinson et al., 2014) – Land use features (Australian Government Productivity Commission 2005Wilkinson et al., 2014; Wilkinson & Remøy, 2018)	6
Political	– Political mandates (Pickerill & Armitage, 2009) – Local government support (Aplin, 2002; Yung & Chan, 2012)	2

Physical	– Complex process (Finch & Kurul, 2007; Wilkinson et al., 2014) – Finding new elements or components through adaptive reuse process	2
Locational	– Being on heritage precinct – Locating on city centres and/or valuable land (Australian Government Productivity Commission, 2005	2
Technical	– Providing disability access (Shipley et al., 2006) – Proving required performance standard and preserving the visual quality (Shipley et al., 2006) – Installation and upgrade of mechanical and electrical systems (Conejos et al., 2016; Historic England, 2020) – Lack of experience and knowledge (Shipley et al., 2006; Koslow, 2010) – Sourcing original materials and components (Douglas, 2006; Bullen, 2007; Bullen & Love, 2011) – Inflexible building (Bullen & Love, 2011)	6

9.4 Previous adaptation and adaptive reuse assessment models

A model considering categories affecting decision-making in relation to implementing adaptation was presented by Wilkinson (2011). This model, called 'Preliminary Assessment Adaptation Model (PAAM)', follows a sequence of conditions which need to be met for an existing building to be considered suitable for adaptation (Figure 9.1).

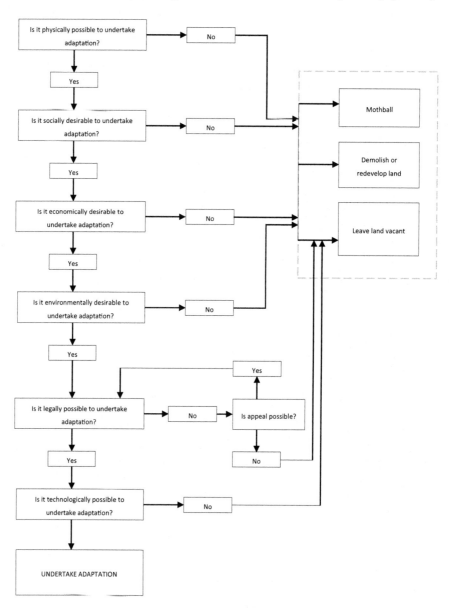

Figure 9.1 Preliminary Assessment Adaptation Model (PAAM)

(Source: Wilkinson, 2011)

PAAM was designed based on the previous adaptation assessment and decision-making tools and models of Chudley (1981), Kincaid (2000), Langston and Shen (2007) and ARUP (2008). PAAM is proposed for the assessment of a building at a certain point in time, as the situation of the building may change over time (Wilkinson, 2011). PAAM can provide a comprehensive basis for the proposed model; however, other models such as Kincaid's (2000) and ARUP's (2008) do not cover previously identified categories, mostly focusing on the levels of adaptation.

Wilkinson et al. (2014) used a large comprehensive adaptation database to glean important attributes of adaptation projects to develop the PAAM. As stated by Wilkinson (2011), the PAAM is mainly derived from Chudley's model (Chudley, 1981) (Figure 9.2); however, there are some differences in the sequencing of stages.

Economic consideration of an adaptation project is the first point in Chudley's model (Chudley, 1981), followed by regulatory requirements, social considerations, aesthetic issues and the required time for a project. According to Wilkinson et al. (2014), Chudley's (1981) model includes almost all identified categories by researchers (Kincaid, 2000; Langston & Shen, 2007; ARUP,

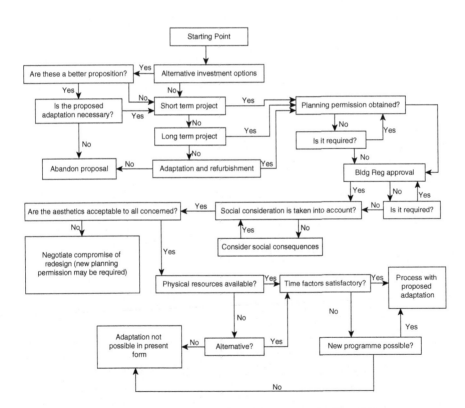

Figure 9.2 Model of decision-making in building adaptation, Chudley, 1981
(Source: Wilkinson, 2011)

2008), with the exception of environmental and technical categories. Furthermore, Chudley did not consider the sequence of stages, to show which categories are more important than others. A further weakness is that, although Chudley's model is easy to follow, it has never been tested in practice. Accordingly, Wilkinson (2011) re-ordered the Chudley model, and considered environmental and technical factors, then tested the PAAM on several real-world case studies. As such, the PAAM, which is also derived from analysis of the preceding models (Chudley, 1981; Kincaid, 2000; Langston & Shen, 2007; Arup, 2008), covers previously identified categories that need to be addressed in applying adaptation for an existing building but lacks the political category. In their models, Chudley (1981) and Wilkinson (2011) did not consider political factors, as well as the heritage value and authentic features, making their models less applicable to heritage buildings.

Wilkinson et al. (2014) stated that PAAM could be used by non-experts to make a preliminary assessment of a building for minor adaptation. However, this model may not be applicable for all existing buildings, as specific features of each building may present a different sequencing, thus changing the importance of categories. Wilkinson et al. (2014) stated a weakness of the PAAM is the lack of evidence to validate the sequence of stages. An important further point is that heritage building characteristics require following a particular sequencing of the identified categories, regardless of their use. Furthermore, both Chudley's (1981) and Wilkinson's (2011) models comprise stages that are based on identified categories which are considered as a whole, but here, identified categories are considered in terms of drivers and challenges. The main focus of the new research model proposed by Yazdani Mehr and Wilkinson (2021) is on challenges to the adaptive reuse of heritage buildings which need to be addressed when applying adaptation.

In the PAAM, the first consideration is given to the physical suitability of a building for adaptation. Accordingly, if a building fails to meet this requirement, it cannot be considered for adaptation even if other conditions are met. In PAAM, the physical condition of an existing building plays an important role in adaptation, whilst technological condition is considered to be the least important category. However, prioritisation of Wilkinson's (2011) model may differ for heritage buildings due to the building's specific value-representative characteristics, and the given obligation to their conservation. For heritage buildings, adaptive reuse is considered as a strategy to conserve buildings for present and future generations, thus the priorities for adaptation are focused on the preservation of heritage values, making this prioritisation different from other existing buildings (Aplin, 2002; Jokilehto, 2007; ICOMOS, 2013).

As demonstrated in Figure 9.1, failure to meet the legal and technological requirements leads to leaving a building/property vacant, whilst for other conditions, a building may be mothballed, demolished or redeveloped (Wilkinson, 2011). However, all three decisions could be applicable to existing buildings if they fail to meet suitability in each category. Accordingly, the dash-line has been added (Figure 9.3), reflecting this new perspective on the model.

9.5 A new model for adaptive reuse assessment of heritage buildings

Based on the analysis of the drivers and challenges to the adaptive reuse of heritage buildings (Tables 9.1 and 9.2), a new model is proposed (Figure 9.3). This new model of the assessment and decision-making process for the adaptive reuse of heritage buildings extends Wilkinson's (2011) model to include the political category, heritage values and authentic features, whilst also giving new prioritisation to Wilkinson's existing categories. These new additions are represented in yellow in Figure 9.3.

Social factors play an important role in the adaptive reuse of heritage buildings since these buildings are usually protected for the long-term benefit of the community. Thus, a heritage building is always considered socially desirable, regardless of its current use or its reuse potential. This is confirmed by Lehne (2008)

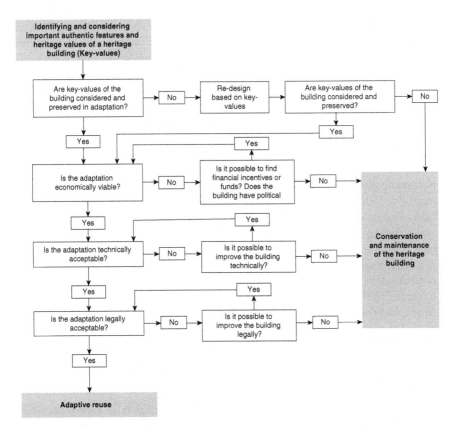

Figure 9.3 Adaptive reuse assessment of heritage buildings

(Source: Yazdani Mehr, 2019b)

who stated that the connected feeling to heritage buildings is the driving force for their protection rather than the historic or aesthetic values of the building. This proposed model (Figure 9.3) starts with identifying and considering important authentic features and heritage values of a building (key values). Depending on the adaptive reuse assessment of a heritage building in each stage, two outcomes are considered: either conservation/maintenance of heritage buildings or adaptive reuse. All stages have a pathway, along which their applicability for adaptive reuse or conservation/maintenance is assessed, and which may loop back depending on meeting the requirements on each pathway. Identifying the key values of heritage buildings might be challenging, due to different stakeholders, perceptions and definitions of values (Pearson & Sullivan, 1995; Irons & Armitage, 2011; ICOMOS, 2013; Wilkinson et al., 2014). Adaptive reuse of a heritage building should aim at preserving these values. A failure to preserve these key values will result in the conservation or maintenance of a heritage building in its existing condition.

Regarding heritage buildings, a successful adaptive reuse project must appropriately address all challenges to adaptation and simultaneously preserve heritage values. Drivers are considered to be benefits which further promote adaptation. For the proposed model, prioritisation is based on the sequence of challenges which received consideration in research. Table 9.2 shows most researchers have mainly considered economic, technical and legal challenges, whilst political, locational, physical and environmental challenges have been less considered. Table 9.2 does not show the importance of the challenges; however, it shows that in the adaptive reuse of heritage buildings, stakeholders are mainly concerned to address the economic, technical and legal challenges in the adaptive reuse of heritage buildings. Accordingly, the next step is the economic viability of the adaptation of heritage buildings, following the technical and legal considerations. The political category, to some extent, overlaps with the economic category in terms of government support and political ownership, and thus given equal consideration in the economic category. Failure in meeting these requirements leads to the conservation and maintenance of a heritage building in its existing condition. Apart from the first stage, the sequence of stages in the proposed model does not necessarily show the importance of categories.

In the proposed model, locational, physical and environmental categories are not considered as these challenges have received the least consideration in previous research. In the proposed model (Figure 9.3), all categories are considered as a whole, and the sequence of stages is based on the frequency of literature reference to challenges in the adaptive reuse of heritage buildings.

9.6 Case studies of sustainable and resilient retrofits of heritage buildings

The following city halls, located in Queensland, Australia, have experienced sustainable and resilient retrofits over time and are presented as illustrative case studies of the complex issues involved in decision-making around sustainable resilient retrofits of heritage buildings in practice.

9.6.1 Johnston Shire Hall

The Johnstone Shire Hall is located in Innisfail, between Cairns and Townsville (See Plate 9.1). The building was designed and constructed in the Art Deco style in 1938. The building has acted as an emergency centre during cyclones and floods and has been used for a variety of functions, such as a picture theatre, for art exhibitions, concerts, dances and school functions. Although the building has undergone adaptation over time, it is still a place for the local government and includes a public hall for the community (Lovell, 2006).

The Hall was constructed in four main levels – basement, administration spaces on the second and third levels and an auditorium on the top level (Johnson, 2009). The building was registered on the Queensland heritage list in 1995 due to its heritage values (Queensland Government, 2016). The building has experienced different adaptations during its lifecycle. The primary type of adaptation work previously undertaken on the building has related to technical and service upgrades, which is the result of the advances in technology and changes that the community demands over time. However, in 2006, after Cyclone Larry caused extensive damage to the

Plate 9.1 Johnstone Shire Hall
(Source: Google Maps, 2018)

building, it experienced massive adaptation. During this adaptation, emphasis was placed on restoring the tangible and intangible values of the building in terms of its heritage values and architectural details, and a technical upgrade of the building.

Even though the adaptation of the building after the cyclone was mainly focused on repair and refurbishment, the council took the opportunity to upgrade the building in respect of technical and environmentally sustainability. Fire safety and air-conditioning were addressed in the adaptation of the building to bring it up to modern standards (Paterson, 2008). The hall on the second floor was upgraded in terms of sound, lighting, bar, kitchen and a lift (Russo, 2008). An additional upgrade included the installation of a stage fly system used to lift scenery from the stage, dressing room fit out, stage improvements and a 'bio box' – a dedicated area/ room used by Technicians to operate and control audio visual equipment for an event. The council planned to upgrade the Hall based on contemporary standards in order to provide the opportunity to attract 'quality touring performers' and local artists (Guy, 2007:4). The Hall was upgraded with state-of-the-art lighting, sound and vision (Lightfoot, 2009; Webster, 2009). Accordingly, the building experienced sustainable/resilient retrofits. Technical upgrade of the building made it an appealing venue for the community and artists, and thus has a financial benefit for the council through the renting of the hall.

9.6.2 Pine Rivers Shire Hall

Pine Rivers Shire is located in South East Queensland, north-west of Brisbane (University of Queensland, 2014). In 1889, a rectangular single storey timber building was constructed in Strathpine as the first Pine Rivers Divisional Board's hall (Queensland Government, 2016) (See Plate 9.2). The building's plans were prepared by the Chairman of the Board, Mr. Henry Ireland (Conway & Oliver, 1978). The efforts of the local council to construct a divisional hall represents their desire to have a fit for purpose building. In 1979, the Shire Hall was listed as a heritage building by the National Trust of Australia (Queensland).

The Hall was used for council business and social activities (University of Queensland, 2014), has undergone three extensive extensions and was used for several functions over time. Adaptation work was mainly focused on the extension of the building to provide space for the local government and community. Internal adaptations were carried out to meet user needs and to preserve the heritage values of the building, being as environmentally sustainable as possible. Accordingly, the building underwent sustainable, resilient retrofits.

Advances in technology and changes in user demand called for a comfortable internal environment. Radford (2017) noted the hall is an old timber building with inappropriate insulation resulting in overheating of the internal space during the summer months, necessitating the installation of an air-conditioning system in 1992. Furthermore, according to the report of the National Trust of Queensland (1978), the Shire Hall needed some levels of technical upgrade such as checking the roof structure, footings and foundations, and electrical rewiring, in order to make the building viable and usable.

Plate 9.2 Pine Rivers Shire Hall
(Source: Yazdani-Mehr, 2017)

9.6.3 *Toowoomba City Hall*

Toowoomba City Hall is located on Ruthven Street in Darling Downs in South West Queensland (See Plate 9.3). It is a two-storeyed masonry building with a central tower, constructed in 1900 with rendered classical detailing (The Queensland Government, 2016). The building was designed in the Federation architectural style and has undergone different adaptations over time. The building has been used for different functions such as educational, cultural and social events. During 1963, the building was completely used for educational activities and as an art gallery until 1994 when the council returned to the building (The University of Queensland, 2014). The main function of the building has almost always remained the same, being the Toowoomba City Hall.

The building experienced major adaptations which were implemented mostly before the building was registered on the heritage list. Extensions to the building were constructed during the 1930s and 1950s, with the aim of gaining more space and upgrading the building to enable new uses, thus addressing functional obsolescence of the building. These additions were successfully integrated into the original concept and were compatible with the original building through applying, as much as possible, the same architectural style and materials, thus creating a unity between the original structure and extensions.

Plate 9.3 Toowoomba City Hall
(Source: Yazdani-Mehr, 2017)

The building has experienced sustainable and resilient retrofits. Improving the internal environmental condition of the building in response to changes in user demands resulted in the introduction of a mechanical ventilation system into the building in circa 1995. Acoustic and thermal insulation were installed to enhance the internal environment of the theatre for the audiences, saving on energy costs. At the present time, the main building is not adequately soundproofed and insulated, which will necessitate further sustainable adaptation of the building. Adam (2016) noted the adaptation of the building is aimed at making the building usable, modern and as future proof as possible.

9.7 Inherent sustainability in heritage buildings

Heritage buildings tend to be inherently environmentally sustainable, with high levels of embodied energy. Accordingly as stated in the report prepared by Historic England (2020), the reuse or conservation of heritage buildings reduces the pressure on natural resources through reducing the need for raw materials, thus contributing to green sustainable future.

Buildings from pre-industrial periods are likely to contain more natural, low-energy, less chemical-based material. However, they predate concerns about energy and water use. Therefore, the consumption of water and energy may be higher comparative to some modern counterparts, although this is highly dependent on a number of factors including occupant behaviours. However, opportunities do exist to optimise water and energy use in heritage buildings, while preserving their heritage values and significance.

Many older buildings were designed and built to work with local climate. The traditional Queenslander house is a good example, with a design that allows air flow below the building to keep it cooler. In other areas, high thermal mass kept buildings cool in summer and warm in winter (Osborne, 2014).

Although heritage buildings vary case-by-case depending on their characteristic features such as size, scale, design, material, geographical location and structure. As noted by international scholars (Douglas, 2006; Clark, 2008; Wilkinson et al., 2014; Mehr et al., 2018; Historic England, 2020; Organ et al., 2020), the preservation of heritage buildings contributes to the reduced carbon emissions, landfill waste, demolition energy use and new construction globally. Adaptive reuse concepts, renovations for less energy use, maintenance and type of use also affect heritage building sustainability. Applying circular economy principles to renovations can improve sustainability further. By considering energy and material use, building quality and adaptability, sustainability issues can be discussed and the integral sustainability of heritage buildings better accounted for in retrofit projects.

9.8 It is not easy being 'green'

Heritage and sustainable development are intimately linked. Understanding heritage helps us better comprehend cultural and social systems. It is vital to understand the philosophical definitions of terms.

Sustainability is an important tool for heritage preservation. Another consideration is time, which includes the fact that changing, evolving uses can add to heritage. Many buildings we classify as 'heritage' have changed greatly over time. For instance, the Tower of London has in its 900-year history been a castle, home, museum, prison and tourist attraction. If we are to maintain social and economic use, sometimes changes and new uses need to be accepted. In these cases, reversibility of repairs is an effective approach.

To sum up: it is complex. Buildings are unique and need to be assessed on various levels:

- What is the heritage value?
- What work is proposed?
- What are existing levels of sustainability?
- Can measures be incorporated in a way that is reversible and does not damage the original building fabric?

Preserving heritage and sustainable development are both important goals. Conflicts need to be identified, assessed, documented and managed to achieve an optimum balance.

9.9 Conclusion

The importance of heritage buildings and their adaptation are discussed in this chapter. Regulations, methods and models related to heritage buildings have been presented to address the protection and adaptation of these buildings. However, a comprehensive review of published works related to adaptive reuse showed that a new model is specifically required for adaptive reuse assessment of heritage buildings. Analysis of previous assessment and decision-making models revealed that political factors as well as heritage values and authentic features of heritage buildings have not been previously considered. For heritage buildings, a successful adaptive reuse project must appropriately address all the challenges to adaptation, whilst simultaneously preserving heritage values. Furthermore, for buildings which fail to address challenges across a range of categories, previous models have considered different options such as demolition, leaving vacant and mothballing for existing buildings. However, heritage buildings must be safeguarded, no matter whether they address each category or not. Another weakness of previous models relates to the fact that at each stage of the adaptive reuse of existing buildings, drivers have not been sufficiently distinguished from challenges. Instead these have been considered together. By considering these factors, this new model starts with the identification of the heritage values and authentic features of a heritage building. For this model, consideration is given mainly to the challenges to the adaptive reuse of heritage buildings, which must be thoroughly addressed to successfully deliver an adaptive reuse project. The new model has the capability of being used in practice for the assessment of the adaptability of heritage buildings globally. A key advantage of the model is its simplicity, since it can be used by stakeholders in adaptive reuse of heritage buildings regardless of their level of expertise. Applying this new model in practice could further strengthen its validity.

This chapter addressed questions related to the definitions and applications of genius loci and authenticity which play important roles in the adaptation of heritage buildings. Genius loci was defined and interpreted as sense of, and spirit of, place. Authenticity was described as being original, representing specific characteristics of a heritage building. Genius loci and authenticity cover tangible and intangible values of a heritage building. Thus, different features and qualities in a heritage building may indicate the genius loci and authenticity, and may imply some level of complications in their preservation.

Authentic features in terms of design, form, workmanship, context, setting and structure contribute to the genius loci. However, the preservation of authentic features and values of heritage buildings may merely result in preservation of the spirit of place. Russell et al. (2011) stated that sense of place has particular meaning and significance for each individual. Graham et al. (2009) acknowledged that people are a crucial indicator of sense of place. These statements imply sense of place is subjective, and individuals may have different feelings related to a place, which create different interpretations. Accordingly, the sense of place differs between communities, individuals and cultures, and thus its preservation is challenging and needs community involvement and communication. However, according to ICOMOS (2008), spirit of place changes over time, and from one culture to another. Petzet (2008:6) stated that 'the spirit of place is transmitted by living people in their every-day [sic] experience and therefore depends entirely on them for its survival'. Thus, the preservation of spirit of place needs the involvement of governments, stakeholders, multidisciplinary experts and local communities, due to its complex and multiform nature. It is a key component of resilient sustainable heritage retrofit or adaptive reuse.

Though these concepts are usually studied separately, this chapter shows they are interconnected. Further, genius loci and authenticity have broad interpretations, which makes their preservation challenging. While this chapter has provided a comprehensive explanation related to these important concepts, there is still debate regarding genius loci and authenticity in practice. Some experts believe that the original features and values of heritage buildings are representative of authenticity and genius loci (Burman, 2008; Del Río Carrasco, 2008; Karsten, 2017). Others believe that all adaptation works carried out over time on heritage buildings add to the authenticity and genius loci, and thus need to be preserved (Jokilehto, 2007; Petzet, 2008; Machat, 2010; Bridgwood & Lennie, 2013; Plevoets, 2014; Ward, 2015). Del Río Carrasco (2008) notes different factors need to be considered in the preservation of authenticity, and the historical basis of a heritage building must be respected. This shows the importance of originality and historicity in the preservation of authenticity and genius loci.

Authenticity remains an essential criterion for UNESCO (1979) in relation to cultural property and the World Heritage List. All features that are listed as 'authentic' in relation to World Heritage directly or indirectly contribute to sense and spirit of place. As such, expert skills and close involvement with the project are needed for the preservation of authenticity and genius loci in the adaptive reuse of heritage buildings. It is the responsibility of engaged parties to thoughtfully consider the authentic features, qualities and values of a heritage building in every adaptation work in order to preserve both authenticity and genius loci.

Table 9.3 Checklist for assessment of genius loci and authenticity in adaptive reuse of heritage buildings

Questions	Yes	No	Comments
1. Are all heritage values of the building identified and documented?			
2. Are the form and design original or representative of a specific period of time?			
3. Does the building function based on its original use? Does the original use of the building represent an important function?			
4. Does the building represent an important historical era or event?			
5. Does a building belong to an important individual or organisation?			
6. Are the construction materials and substances original or representative of a specific period of time?			
7. Is the building located in its original setting? Is the location of the building important?			
8. Does the building represent specific techniques and traditions internally and externally?			
9. Do past adaptation works represent a specific era or function? Do these adaptation works add to the values of the building?			

(Source: Yazdani Mehr, 2019b).

Based on different definitions of and perspectives on genius loci and authenticity, Table 9.3 provides a checklist of questions in order to guide the preservation of the genius loci and authenticity of a building. This checklist can easily be applied by those involved in the adaptive reuse of heritage buildings and contributes to knowledge and practice.

References

Adam, A., 2016. Adaptive reuse of Toowoomba city hall (PhD S. Y. Mehr), Griffith University, Gold Coast, Australia.

Aplin, G., 2002. *Heritage: Identification, Conservation, and Management.* Oxford University Press, Oxford.

Armitage, L. and Irons, J., 2013. The values of built heritage. *Property Management, 31* (3), pp. 246–259.

Arup, F. B. S. S., 2008. *A Toolbox for Re-Energising Tired Assets.* Arup and Property Council of Australia (PCA), Victorian Division, Australia.

Australian Government Productivity Commission, 2005. *Conservation of Australia's Historic Heritage Places.* Productivity Commission, Melbourne.

Azizi, N. Z. M., Razak, A. A., Din, M. A. M. and Nasir, N. M., 2016. Recurring issues in historic building conservation. *Procedia-Social and Behavioral Sciences, 222*, pp. 587–595.

Ball, R., 2002. Re use potential and vacant industrial premises: Revisiting the regeneration issue in Stoke-on-Trent. *Journal of Property Research, 19*(2), pp. 93–110.

Binder, M. L., 2003. *Adaptive Reuse and Sustainable Design: A Holistic Approach for Abandoned Industrial Buildings.* University of Cincinnati, Cincinatti, OH.

Boito, C. and Birignani, C., 2009. Restoration in architecture: First dialogue. *Future Anterior, 6*(1), pp. 68–83.

Bold, J., Larkham, P. and Pickard, R., 2017. *Authentic Reconstruction: Authenticity, Architecture and the Built Heritage.* Bloomsbury Publishing, London.

Bridgwood, B. and Lennie, L., 2013. *History, Performance and Conservation.* Taylor & Francis, New York.

Brooker, G. and Stone, S., 2004. *Rereadings: Interior Architecture and the Design Principles of Remodelling Existing Buildings.* RIBA Enterprises, London.

Brown, S., 2004. How to extract cash from old bricks. *The Estates Gazette, 4*(436), pp. 112–113.

Bullen, P. and Love, P., 2011. Factors influencing the adaptive re-use of buildings. *Journal of Engineering, Design and Technology, 9*(1), pp. 32–46.

Bullen, P. A., 2004, September. Sustainable adaptive reuse of the existing building stock in Western Australia. In *20th Annual ARCOM Conference* (Vol. 1, No. 3).

Bullen, P. A., 2007. Adaptive reuse and sustainability of commercial buildings. *Facilities, 25*(1/2), pp. 20–31.

Bullen, P. A. and Love, P. E., 2009. Residential regeneration and adaptive reuse: Learning from the experiences of Los Angeles. *Structural Survey, 27*(5), pp. 351–360.

Bullen, P. A. and Love, P. E., 2011a. A new future for the past: A model for adaptive reuse decision making. *Built Environment Project and Asset Management, 1*(1), pp. 32–44.

Bullen, P. A. and Love, P. E., 2011b. Adaptive reuse of heritage buildings. *Structural Survey, 29*(5), pp. 411–421.

Burman, P., 2008. Ruskin's children: John Ruskin (1819–1900), the good steward, and his influence today. Conservation and preservation: Interactions between theory and practice: In memoriam Alois Riegl (1858–1905). Proceedings of the International Conference of the ICOMOS International Scientific Committee for the Theory and the Philosophy of Conservation, Polistampa.

Ch'ng, K., 2010. *The Beneficial Past: Promoting Adaptive Reuse as a Beneficial Design Method for East and South-East Asia.*

Chudley, R., 1981. *The Maintenance and Adaptation of Buildings.* Longman, London.

Clark, K., 2008. Only connect – sustainable development and cultural heritage. *The Heritage Reader*, pp. 82–98.

Conejos, S., 2013. *Designing for Future Building Adaptive Reuse.* Bond University, Brisbane, Queensland.

Conejos, S., Langston, C., Chan, E. H. and Chew, M. Y., 2016. Governance of heritage buildings: Australian regulatory barriers to adaptive reuse. *Building Research & Information, 44*(5–6), pp. 507–519.

Conejos, S., Langston, C. and Smith, J., 2013. AdaptSTAR model: A climate-friendly strategy to promote built environment sustainability. *Habitat International, 37*, 95–103.

Conway, T. and Oliver, R., 1978. *The Old Shire Hall Strathpine. T. N. T. o. Queensland.* The National Trust of Queensland, Australia.

Del Río Carrasco, J. M., 2008. Values of heritage in the religious and cultural tradition of Christianity: The concept of authenticity. *Values and Criteria in Heritage Conservation, Polistampa*, pp. 1000–1023.

Department of Planning Sydney, R., 1995. Heritage assessment guidelines. *Crown Copyright*.

Douglas, J., 2006. *Building Adaptation*. Routledge, London.

Finch, E. and Kurul, E., 2007. A qualitative approach to exploring adaptive re-use processes. *Facilities*, 25(13/14), pp. 554–570.

Gallie, W. B., 1955. Essentially contested concepts. Proceedings of the Aristotelian Society, JSTOR.

Graham, H., Mason, R. and Newman, A. 2009. Historic environment, sense of place and social capital. Report commissioned by English Heritage.

Guy, L., 2007. *Shire to Spend $1.25m to Complete Hall Upgrade*. Innisfail Advocate. Queensland, Australia, Innisfail Advocate.

Historic England, 2020. *Heritage Counts: Know Your Home, Know Your Carbon – Reducing Carbon Emissions in Traditional Homes*. Historic England on behalf of the Historic Environment Forum. Available from https://historicengland.org.uk/research/heritage-counts/2020-know-your-carbon/reducing-carbon-emissions-in-traditional-homes/. Accessed 25 October 2021.

Holden, G. 2012. Authentic experience and minor place-making. Paper presented at the Designing Place. International Urban Design Conference, Nottingham.

Hussein, J., 2017. *Conservation of Cultural Built Heritage: An Investigation of Stakeholder Perceptions in Australia and Tanzania*. Doctor of Philosophy, Bond University, Brisbane, Queensland.

International Council on Monuments and Sites (ICOMOS), 2004. *International Charters for Conservation and Restoration*, introduction by Michael Petzet (p. 21). ICOMOS, Paris.

ICOMOS, A., 2008. Quebec declaration of the preservation of the spirit of place. *International Journal of Cultural Property*.

ICOMOS, A., 2013. *The Burra Charter: The Australia ICOMOS Charter for Places of Cultural Significance*.

ICOMOS, N., 1994. *The Nara Document on Authenticity*. ICOMOS.

Irons, J. and Armitage, L., 2011. The value of built heritage: Community, economy and environment. *Pacific Rim Property Research Journal*, 17(4), pp. 614–633.

Jayantha, W. M. and Yung, E. H. K., 2018. Effect of revitalisation of historic buildings on retail shop values in urban renewal: An empirical analysis. *Sustainability*, 10(5), p. 1418.

Johnson, M., 2009. *Renovation of Shire Hall; Official Opening*. Brisbane, Queensland.

Jokilehto, J., 2007. *History of Architectural Conservation*. Routledge, London.

Jokilehto, J., 2008. The idea of conservation: An Overview. Conservation and preservation: Interactions between theory and practice: In memoriam Alois Riegl (1858–1905). Proceedings of the International Conference of the ICOMOS International Scientific Committee for the Theory and the Philosophy of Conservation, Polistampa.

Karsten, I. A., 2017. Reconstruction of historic monuments in Poland after the Second World War – the case of Warsaw. *Authentic Reconstruction: Authenticity, Architecture and the Built Heritage*, p. 47.

Kepczynska-Walczak, A. and Walczak, B. M., 2013. Visualising "genius loci" of built heritage. Envisioning Architecture: Design, Evaluation, Communication, Proceedings of the 11th conference of the European Architectural Envisioning Association. Edizioni Nuova Cultura, Rome.

Kincaid, D., 2000. Adaptability potentials for buildings and infrastructure in sustainable cities. *Facilities*, 18(3/4), pp. 155–161.

Koslow, J., 2010. Opportunities and challenges in whole-building retrofits (Doctoral dissertation).

Kwanda, T. J. P. E., 2015. Authenticity principle in conservation of de Javasche Bank of Surabaya: Materials, substance and form, *Procedia Engineering, 125,* 675–684.

Laenen, M., 2008. Reflections on heritage values. *Values and Criteria in Heritage Conservation, Polistampa,* pp. 1000–1009.

Langdon, D., 2008. *Opportunities for Existing Buildings: Deep Emission Cuts.* Australia Davids Langdon, Melbourne, Victoria.

Langston, C., 2008, October. The sustainability implications of building adaptive reuse. CRIOCM 2008 International Research Symposium on Advancement of Construction Management and Real Estate, Beijing, China.

Langston, C. and Shen, L. Y., 2007. Application of the adaptive reuse potential model in Hong Kong: A case study of Lui Seng Chun. *International Journal of Strategic Property Management, 11*(4), pp. 193–207.

Latham, D., 2000. *Creative Reuse of Buildings: Volume One.* Routledge. London.

Lehne, A., 2008. Georg Dehio, Alois Riegl, Max Dvorák-a Threshold in theory development. Conservation and preservation: Interactions between theory and practice: In memoriam Alois Riegl (1858–1905). Proceedings of the International Conference of the ICOMOS International Scientific Committee for the Theory and the Philosophy of Conservation, Polistampa.

Lightfoot, J., 2009. New shire hall comes with $17m price tag. *The Cairns Post.* Queensland, Australia, The Cairns Post.

Lovell, A., 2006. *Johnstone Shire Hall, Innisfail. A. Brisbane.* Allom Lovell Architects Brisbane, Australia.

Machat, C., 2010. The vernacular between theory and practice. Conservation and preservation: Interactions between theory and practice: In memoriam Alois Riegl (1858–1905). Proceedings of the International Conference of the ICOMOS International Scientific Committee for the Theory and the Philosophy of Conservation, Polistampa.

Mehr, S. Y. and Wilkinson, S., 2018. Technical issues and energy efficient adaptive reuse of heritage listed city halls in Queensland Australia. *International Journal of Building Pathology and Adaptation.*

New South Wales Heritage Office & Royal Australian Institute of Architects, New South Wales Chapter & Heritage Council of New South Wales, 2008. *New Uses for Heritage Places: Guidelines for the Adaptation of Historic Buildings and Sites.* Heritage Office of NSW, NSW Dept. of Planning, Parramatta, New South Wales.

O'Donnell, C., 2004. Getting serious about green dollars. *Property Australia, 18*(4), pp. 1–2.

Organ, S., Lamond, J., Drewniak, D. and Wood, M., 2020. Carbon reduction scenarios in the built historic environment. *Historic England.*

Osborne, L., 2014. Sublime design: The Queenslander. *The Conservation.* Available from https://theconversation.com/sublime-design-the-queenslander-27225.

Paterson, D., 2008. *More Dollars for Hall. Innisfail Advocate.* North Queensland, Australia, Innisfail advocate.

Pearson, M. and Sullivan, S., 1995. *Looking after Heritage Places: The Basics of Heritage Planning for Managers, Landowners and Administrators.* Melbourne University Press, Melbourne.

Penna, K. N., 2014. Sustainable conservation: Transforming conceptual dualism in harmony for the safeguard of historic cities in developing countries. Proceedings of the 4th International Conference on Heritage and Sustainable Development, Heritage.

Petzet, M., 2008. *Genius Loci – the Spirit of Monuments and Sites.*

Pickerill, T. and Armitage, L., 2009. The management of built heritage: A comparative review of policies and practice in Western Europe, North America and Australia. *Management, 2009,* pp. 01–21.

Plevoets, B., 2014. Retail-Reuse: An interior view on adaptive reuse of buildings (PhD). Hasselt University Press, Hasselt, Belgium.

The Queensland Government, R., 2016. *Toowoomba City Hall. H. Places.* Queensland Government, Australia.

Queensland Government, R., 2016. *Johnstone Shire Hall.* Queensland Heritage Register, Australia.

Radford, M., 2017. Adaptive reuse of pine rivers shire hall (PhD). S. Y. Mehr. Australia, Griffith University.

Reyers, J. and Mansfield, J., 2001. The assessment of risk in conservation refurbishment projects. *Structural Survey, 19*(5), pp. 238–244.

Ruggles, D. F. and Silverman, H., 2009. From tangible to intangible heritage. In *Intangible Heritage Embodied* (pp. 1–14). Springer, New York.

Russell, H., Smith, A. and Leverton, P., 2011. *Sustaining Cultural Identity and a Sense of Place: New Wine in Old Bottles or Old Wine in New Bottles?* College of Estate Management, Reading.

Russo, F., 2008. *Old Girl to Miss Birthday. Innisfail Advocate.* North Queensland, Australia, Innisfail Advocate.

Schofield, J., 2008. Heritage management, theory and practice. *The Heritage Reader,* pp. 15–30.

Scott, F., 2007. *On Altering Architecture.* Routledge, London.

Shinbira, I. A., 2012. Conservation of the urban heritage to conserve the sense of place, a case study Misurata City, Libya. 2nd International Conference-Workshop on Sustainable Architecture and Urban Design.

Shipley, R., Utz, S. and Parsons, M., 2006. Does adaptive reuse pay? A study of the business of building renovation in Ontario, Canada. *International Journal of Heritage Studies, 12*(6), pp. 505–520.

Slocombe, M., no date. The SPAB approach to the conservation and care of old buildings. Available from www.spab.org.uk/sites/default/files/documents/MainSociety/Campaigning/SPAB%20Approach.pdf. Accessed 18 January 2022.

Society for the Protection of Ancient Buildings, SPAB, 2017. The SPAB manifesto. Available from www.spab.org.uk/about-us/spab-manifesto#:~:text=It%20is%20for%20all%20these,meant%20for%20support%20or%20covering%2C. Accessed 18 January 2022.

Tomaszewski, A., 2007. Values and criteria in heritage conservation. Proceedings of the International Conference of ICOMOS, ICCROM, Fondazione Romualdo Del Bianco.

UNESCO, 1979. Policies regarding credibility of the world heritage list – decision of the world heritage committee CONF 003 XI.35. Available from https://whc.unesco.org/en/compendium/269. Accessed 15 January 2022.

The University of Queensland, R., 2014. *Toowoomba City Hall. A. E-Heritage.* The University of Queensland, Australia.

University of Queensland, R., 2014. *Pine Rivers Shire Hall (former). A. E-Heritage.* The University of Queensland, Australia.

Venice Charter, R., 1964. *International Charter for the Conservation and Restoration of Monuments and Sites.* Venice, Italy.

Ward, S., 2015. *Authenticating Adaptation.* Carleton University, Ottawa.

Webster, A., 2009. Hall returns to public service. *The Cairns Post.* Queensland, Australia, The Cairns Post.

Wilkinson, S., 2011. *The Relationship between Building Adaptation and Property Attributes.* Deakin University, Australia.

Wilkinson, S. and Remoy, H., 2017. Adaptive reuse of Sydney offices and sustainability. *Sustainable Buildings*, 2, p. 6.

Wilkinson, S. J. and Remøy, H., 2018. *Building Urban Resilience Through Change of Use*. John Wiley & Sons, London.

Wilkinson, S., Remøy, H. and Langston, C., 2014. *Sustainable Building Adaptation: Innovations in Decision-Making*. John Wiley & Sons, London.

Willson, G. B. and McIntosh, A. J., 2007. Heritage buildings and tourism: An experiential view. *Journal of Heritage Tourism*, 2(2), pp. 75–93.

Wong, L., 2016. *Adaptive Reuse: Extending the Lives of Buildings*. Birkhäuser.

Yau, Y., Wing Chau, K., Chi Wing Ho, D. and Kei Wong, S., 2008. An empirical study on the positive externality of building refurbishment. *International Journal of Housing Markets and Analysis*, 1(1), pp. 19–32.

Yazdani Mehr, S., 2019a. Analysis of 19th and 20th century conservation key theories in relation to contemporary adaptive reuse of heritage buildings. *Heritage*, 2(1), pp. 920–937.

Yazdani Mehr, S., 2019b. Adaptive reuse of heritage listed city halls in Queensland, Australia. (PhD), Griffith University Queensland Australia. Available from https://research-repository.griffith.edu.au/handle/10072/387684. Accessed 25 January 2021.

Yazdani Mehr, S. and Wilkinson, S., 2021. A model for assessing adaptability in heritage buildings. *International Journal of Conservation Science*.

Yung, E. H. and Chan, E. H., 2012. Implementation challenges to the adaptive reuse of heritage buildings: Towards the goals of sustainable, low carbon cities. *Habitat International*, 36(3), pp. 352–361.

Yung, E. H., Langston, C. and Chan, E. H., 2014. Adaptive reuse of traditional Chinese shophouses in government-led urban renewal projects in Hong Kong. *Cities*, 39, pp. 87–98.

Part 3

Conclusions, the future, and a manifesto for change

10 Conclusions and a manifesto to retrofit for the future

Sara Wilkinson, Sarah Sayce and Gillian Armstrong

10.1 Introduction

This text has explored 'the why' of the need to deliver resilient building retrofits in the 21st century. The first four chapters examined the challenges society faces and why the need for change is paramount. An overview of our existing stock of buildings was presented as well as an explanation of how we understand vacancy in office buildings. Covid-19 has accelerated social and economic change globally since 2020, which makes a case for rising to the challenge more important. The second part of the book explored 'the what' from the perspective of policymakers, focusing on governance approaches to engage stakeholders to adopt resilient building retrofits and retrofit financing. It also explored some of the existing low and high tech solutions to retrofit and repurposing and looked to what happens with the heritage stock and what we can learn from the past.

This last chapter concludes with 'the when' and 'the how' and a manifesto of recommendations for policymakers, educationalists professional bodies and practitioners. Whilst it may be speculative in some ways, the intent is to underscore the conviction that a business-as-usual model no longer works; it argues that the responses to date to building adaptation are too timid – and that this lack of real commitment and drive could be argued, to quote Greta Thunberg: is 'beyond absurd'. The time for action is now.

10.2 The why: challenge and a need to change

10.2.1 The climate crisis: why it matters for the built environment

The first chapter set the scene regarding the causes of global climate change and the social, environmental, and economic impacts as well as the significant contribution of the built environment to global warming. The whole lifecycle of a building, from its location, through to its construction, use, alteration, and demolition has a major influence on climate, through its contribution to carbon emissions (Clayton et al., 2021). A solution highlighted and stated is the need for more sustainable retrofits to mitigate these impacts and to contribute to lessening waste, reducing building-related water and energy consumption, to adopting a circular

DOI: 10.1201/9781003023975-13

economy approach. Despite knowing that action will mitigate climate change, we have collectively chosen, at best very limited action. This is not sufficient. As climate changes impact further, our built environment will become less fit for purpose, as sea-level rises and flooding and inundation increase, for example. We can increase resilience through retrofit and we need to act.

Socio-political impacts identified include species extinction, disruption to food supply chains, and mass migration. Resilient building retrofits can provide habitat for biodiversity, can accommodate food production, and accommodate displaced people. The chapter reviewed the built environment from city to precinct and building scale in respect of services such as transport, water and energy infrastructure, the issues faced, and the changes that can be delivered in retrofit projects.

10.2.2 The philosophy and definition of retrofitting for resilience

The philosophy and definition of retrofitting for resilience chapter defined what is meant by a resilient building and revealed the wide spectrum of understanding. Not only is retrofit a necessity, due to the climate imperative, but it is also a desirable social outcome, conserving as it does, the maintenance of place, memory, and culture. The chapter debated what happens when place and community are lost for example, as in the London Docklands and explored the extent to which retrofitting to preserve – or conserve – the social value of the building and its context is a philosophical, economic, or legislative matter. The ways resilient retrofit differs from any normal cyclical undertaking to bring a building back to standard and the issue of decisions around the redevelop/retrofit decision were examined. The timing of retrofit in the lifecycle was discussed and the spectrum of 'deep' to 'light' retrofit approaches were described. The chapter emphasised the shortfalls in the business-as-usual approach and outlined the radical and drastic changes needed to our current conceptual understanding to deliver the much-needed changes.

10.2.3 An inadequate building stock

The complex and diverse challenges relating to retrofitting existing buildings are environmental and economic, social, and technological, regulatory and political. They can manifest independently and concurrently and can be acute or chronic. The onset of Covid-19 in 2020, for example, is an acute health shock, which may result in a chronic long-term economic impact. Managing the stock for optimum use and performance in these circumstances is challenging. The complexity is further exacerbated by the diverse range of stakeholder groups; each having different drivers and barriers.

Fewer than 5% of European buildings are estimated as 'energy efficient': far less are zero carbon. Given minimum energy standards in Europe predate other countries by two decades plus, stock elsewhere is even less energy efficient. Energy retrofit is a small part of the picture: water conservation, flood protection, and fire resilience are increasingly important. Appropriate materials with low volatile organic compounds (VOCs) and low embodied energy are needed in projects. The challenge with sustainable, resilient retrofits is to balance what is retained and what is replaced and upgraded.

Caution is needed with new materials as disasters, such as the Grenfell Tower fire show, evidence of gaps in current legislation and design knowledge. Health and well-being are paramount and research connects chronic and fatal conditions with pollution and building defects, and buildings being retrofitted need evaluation. The 2020 Covid-19 pandemic revealed a hitherto unknown risk of disease transmission through HVAC systems, with office workers work remotely to reduce risk of virus exposure and transmission. The ongoing discussions about the extent of the return-to-work post pandemic may lead to high vacancy and more growth in our 'inadequate stock'. Finally, there is the mismatch of buildings to the social needs of communities: future shifts in transport, and the need to accommodate people and to feed them locally collectively creates a multifacetted challenge to adapt and improve our existing buildings.

10.2.4 Understanding vacancy in the office stock

Vacancy is an important indicator of the risk of premature obsolescence in existing buildings. However, the process of obsolescence is poorly studied and documented, with many discussions focusing only on buildings that have reached their terminal condition of long-term structural vacancy and low levels of occupancy, if any. This chapter made a case for a more nuanced understanding of the vacancy through robust datasets to inform and shape effective policy to mitigate obsolescence earlier in a building's lifecycle. This chapter focused on vacancy at the city scale, across building stocks, to develop generalisable findings to build resilient cities. It set out a new method, Vacancy Visual Analytic Method (VVAM), as a practical tool for quantifying vacancy by repurposing data already collected. The reuse of secondary data makes the research process as efficient as possible. The chapter focused on findings from applying VVAM to a case city Adelaide, Australia. The results make the case that vacancy understanding offers insights and challenges prevailing views about vacant space and solutions to mitigate vacancy. We have not developed a consensus of different vacancy types useful as indicators of the risk of premature obsolescence. Currently, the focus is predominantly on simplistic aggregated untenanted vacancy rates. VVAM provides a fine-grain untenanted vacancy data for every building across a city. VVAM also quantifies greyspace vacancy, a hitherto hidden vacancy type. Finally, this chapter set out the need to be more critical of the problematisation of vacancy by industry groups, particularly when fine-grain vacancy is not readily available. A resilient city can weather the stresses and acute shocks with greater levels of social equity, shorter periods of economic decline, and reducing premature obsolescence in cities.

10.3 The what: exploring solutions

10.3.1 A governance response: from persuasive to coercive?

Existing building stocks are largely energy inefficient and unsustainable. In order for greater and increasingly faster take-up of retrofitting, it is important to recognise the diverse range of governance interventions available, from coercive building regulations mandating minimum building performance, through positive financial incentives, and market-led voluntary benchmarking and classification tools. Through evaluating the spectrum of interventions and their different jurisdictions, the solution calls for governments to take an 'orchestrating' role in ensuring the wide range of mutually supportive mechanisms are deployed. An overhaul of building regulation is needed to recognise the spectrum of mechanisms to that there is holistic oversight of the range of incentives, involving a combination of coercive and persuasive interventions which effectively target different groups of property owners and users with tailored regulatory and governance interventions. What is interesting is that there is no single solution and that the solution lies in governments acting as the 'orchestrator' recognising the value of each mechanism and the gaps that need to be narrowed. A hybrid solution can be achieved through a broader rethinking of the regulatory and governance system for building retrofits. The approach to rethink acknowledges the relatively recent substantial changes that have occurred in public governance and regulation in a wide variety of policy areas. The changes have resulted in the shift 'from the government to governance' in many developed countries including the US, the UK, and Australia through the privatisation of public service delivery, shifts in deregulation and reregulation, and the embracing of new public management practices. Building regulation is now an ongoing collaboration between government, the private sector, and civil society as it also includes the growth of self-governing and polycentric initiatives by firms, citizens, and municipalities.

10.3.2 Financing retrofits

Financing for retrofit is rapidly evolving to assist owner-occupiers and investment owners in overcoming economic barriers to finance retrofitting activity. Generally, there is no shortage of public financing and incentives for development, and the cost of borrowing is currently low. Through looking at Europe, which has the oldest building stock of all major developed regions, it is argued that there is the need to direct substantially more of the available finance and investment towards building retrofits through dedicated policy and market instruments to achieve committed carbon targets. Most finance schemes to date have been geared primarily towards increasing the volume of energy retrofits within the residential sector. This focus is understandable as improvements to residential buildings can have significant impacts on human health. However, without substantial increases to available retrofit finance, for all buildings, retrofit rates will not increase sufficiently. While policy tools such as Energy Performance Certificates (EPCs) have contributed to raising awareness of the need to retrofit, they have not succeeded in sufficiently ramping up renovation rates. However, public subsidised funds can only do so much: private equity and debt funding are key to ensuring a sea change in retrofit rates. The challenge now is how to successfully channel and engage the increase in investment to projects that can provide a substantial climate change mitigation impact. One further challenge is the pressing priority to support new occupational requirements for existing buildings due to control of the Covid-19 pandemic. Sectoral fragmentation and the lack of whole lifecycle and circular economy thinking, as well as a lack of data, has resulted in a lack of infrastructure to assist financing the value chain. Currently, there is no universal, standardised system or protocol in place which aids the channelling of finance investments for maximum impact to mitigate and manage climate change. Systems to facilitate access, storage, update and transfer of building-related data and information in a standardised format are long overdue, as it is transparent data that can nudge changes in behaviours and substantiate business cases for the need for retrofit expenditure to ameliorate building value depreciation.

10.3.3 Technological solutions

The well-being of future generations must be factored into the decisions we make from today. Technological solutions, whether they be high or low tech, need to be critically evaluated and understood in terms of their ability to meet net zero carbon, factoring in the embodied energy and operational energy, when designing retrofit interventions. Though there has been some debate about what constitutes a Net Zero Energy (NZE) retrofit of existing buildings, its conceptual roots stem from research from the 1970s and extends this to a clear desired policy outcome to decarbonise the existing building stock through 'deep renovations' to meet the UN's Sustainable Development Goals, which set out the need to double the global rate of improvement in energy efficiency in order to limit climate change and address fuel poverty and energy poverty. A further extension to NZE buildings is 'energy positive' buildings, which produce and export surplus energy throughout the year. The UK Green Building Council (2020) highlights 17 opportunities for net zero carbon-framed using the RIBA Plan of Works Stages. In addition, ARUP Engineers have set out guiding principles for achieving net zero energy and carbon buildings, from passive building designs, reusing materials and refurbishing buildings, decoupling from fossil fuels, reducing operational energy and using energy demand management, to use on-site renewable technology. The variety of options and the combination of them as a holistic solution are wide-ranging, yet NZE are yet to become common practice. 'Low-' and 'high-technology' have been defined in divergent ways and are continuously evolving – from low-technology materials such as strawbales to bio-based materials such as hempcrete and high-tech integrated systems such as algae building technology and Energiesprong, an approach which involves 3D laser scanning of existing houses to transition houses through whole-building refurbishments at scale through innovative solutions and financing mechanisms. The burgeoning of technological innovations associates lower energy consumption and carbon emissions with other innovative practices such as 'disruptive' technologies to create new markets and sustainable business models. Given that a vast majority of our built environment will be here for decades to come, retrofitting buildings offers the greatest chance to decarbonise our built environment.

10.3.4 Repurposing and adaptation

Social and technological change will always affect how buildings are used and managed in ways previously unimagined by the original owners and designers, and change is inevitable. Some changes are unpredictable and fast, known as acute shocks, whereas others are slow and ongoing or chronic. A key premise of this text is that it is imperative to reconsider the case for premature demolition when cities and their buildings are faced with acute shocks or a slow stagnation, as can be seen in projects to repurpose buildings in the rust belt in North America and retail buildings currently impacted by technological shifts enabling rapid uptake in our preference to shop online. What is interesting about the current shocks from Covid-19 since 2020 is the scale at which we see rapid changes in how we use buildings in our cities. This is still having an impact on our existing building stock, and the full outcome is yet to be realised; however, looking at precedents can offer useful explorations to develop future solutions, alongside creative innovative thinking. Repurposing and adapting redundant stock for new uses need to meet revised needs and new demands. Examples explored include urban food production, shared affordable and alternative housing to meet the basic needs of humans and address homelessness, food poverty, and consumption of green belt land. Innovative ideas and policy instruments are offered to highlight the possibilities of repurposing activity around the globe, including redundant stocks in retail, commercial offices, and industrial property sectors. The benefits of repurposing can be felt more widely than the building itself, enhancing existing communities, and adding cultural capital resulting in increased urban vibrancy in cities suffering decline. It seems repurposing buildings not only can add financial value to the immediate building, but increases in value can also be felt on the property values of nearby buildings affected by the blight of redundant industrial buildings. Local governments should therefore consider the conversion of industrial heritage buildings more actively also as a financial instrument in urban redevelopment. And finally, the case is made for the need to repurpose heritage buildings for the wider benefits this brings. Repurposing, adapting, and converting buildings is not a new phenomenon; there are a plethora of conversion examples, as outlined, contributing to today's beloved historical cities and buildings. Repurposing real estate contributes to urban resilience. Together with new concepts of the sharing, circular, and donut economy, repurposing existing buildings can deliver sustainable development and meet social, economic, and environmental needs of current and future shifts to accommodate new use patterns in cities, even when on a global scale not previously seen in recent history.

10.3.5 Heritage: learning from and preserving the past

Heritage buildings tend to be inherently environmentally sustainable, with high levels of embodied energy. They fulfil important demands for cultural experiences and leisure and create benefits for tourism. They are irreplaceable because of the identity and values assigned to them and hold a 'genius loci' which overlays identity of place to a building and its surroundings. The genius loci of heritage buildings has been overlooked in scientific analysis and is a contested term as its complexity offers a multitude of different interpretations. However, preserving the spirit of a place and its authenticity is crucial in developing resilient retrofitting of heritage buildings as social factors play an important role in protecting heritage buildings. A heritage building is always considered socially desirable, regardless of its current use or its reuse potential. Depending on the assessment of a heritage building, two outcomes are considered: either conservation/maintenance of heritage buildings or adaptive reuse. Identifying the key values of heritage buildings might be challenging due to different stakeholders, perceptions, and definitions of values. A failure to preserve these key values may result in a failure of the conservation or maintenance of a heritage building. The social factors which drive the feasibility of adapting heritage buildings range from the perceived heritage value and the impact of proposed new work upon the perceived value. A comprehensive review highlights there is a need for a new model, specifically for adaptive reuse assessment of heritage buildings. Currently, political factors, as well as heritage values and authentic features of heritage buildings, are not well understood in research. Buildings that fail to address challenges, including political factors, may be at risk of unsustainable strategies such as premature demolition or mothballed and left vacant for long periods. However, heritage buildings must be safeguarded, no matter whether they address each category or not. The many categories of consideration, from locational, environmental, social, legal, economic, physical, locational, technical, and political, are intertwined. Through this understanding, a new model is presented, adding a political element to aid decisions in whether to conserve and maintain a heritage building in its current use or convert the building to a new use so as to preserve a building's spirit of place, a complex decision process involving government regulation, stakeholders, multidisciplinary experts, and local communities.

10.4 The when and the how: the need for a manifesto

The urgency and need to act is clear. The brutal truth is that time has run out given our current rates of action and adoption of sustainability measures in the built environment. The longer we wait, the more it will take to catch up. As Greta Thunberg said at the Extinction Rebellion meeting in London in April 2019: '*Humanity is now standing at a crossroads. We must now decide which path we want to take. How do we want the future living conditions for all living species to be like?*' and '*Everything needs to change – and it has to start today*'. There is no longer time to think short, medium and long term; short term is all that is left. But the lesson of the pandemic is that where money, political will, and decisive actions are combined, actions can be undertaken with quick results. Yes, there will be disagreements and there will be a need to coerce, but effective actions can occur.

What are the easy wins? We need to evaluate the 'best bang for the buck' when considering retrofits. What this means is acknowledging that a viable resilient building retrofit option in one location may not be the same for another building in the same location due to differences in design, use, existing materials, and energy and water loads. Equally, a viable resilient building retrofit option in one location will not be the same in another location. For example, location A may be affected by intense and increasing temperatures and heat, whereas another location will be facing rising sea levels and flooding. Priority areas are embodied carbon and reusing existing materials and components, reducing operational energy and water use.

What are the implications for all stakeholders – those who own, those who occupy, professional advisers, educators, and policymakers? Owners need to ensure their properties are fit for purpose; otherwise tenants will prefer to lease other property. Depending on the market, capital values may be impacted negatively if buildings are perceived to be uncomfortable to occupy, expensive to operate, or vulnerable to climate-related events such as flooding or heatwaves. Occupiers will prefer to occupy properties that reflect the values of their business, are attractive to their workforce, and are economic to operate. There are opportunities for built environment professionals to provide advice to clients that demonstrate social, environmental, and economic benefits of owning and occupying sustainable buildings and the options available to retrofit existing stock. Educators play a vital role in preparing current generations with the knowledge and skills to act effectively in retrofitting built environment to increase climate resilience. They can arm people with the skills and knowledge to act and not be daunted by the fear of the unknown. Finally, there are policymakers; their role is to set the standards and provide the framework for us to adopt to ensure a safe, viable environment for all. Their actions are taken on behalf of society at large; their timeframe can span the short to long term. Research has demonstrated in some circumstances mandation is the best option as it sets a minimum standard for all – the danger is we accept the lowest acceptable standard rather than the highest possible standard. We need to guard against this tendency especially prevalent in 'market-led' liberal economies. In other circumstances, there is research showing that the market can deliver good sustainability and retrofit outcomes, provided the incentives are attractive. The challenge, as always, is getting the policy right.

10.5 The manifesto

10.5.1 A manifesto of recommendations for policymakers

The principal desired policy outcome is to decarbonise the existing building stock and transform buildings into ones that are highly energy efficient through 'deep renovations'.

One crucial element in developing an effective urban policy to address premature obsolescence is a robust and critical understanding of the vacancy itself. A lack of fine-grain data to understand where vacancy is sitting also prevents meaningful explorations for policy development, resulting in ad hoc reactive policy to address vacancy and its impacts (Buitelaar et al., 2021). There is an urgent need for local and national governments to collect or access data for understanding vacancy and under-occupation of existing buildings so that sustainable practices to ameliorate vacancy are encouraged through effective policy mechanisms and that policy mechanisms align with the local market conditions, including the types of vacancy and their distribution in different building stocks. Premature demolition of underused buildings can be actively discouraged and effective policy can be developed if there is access to robust data to understand vacancy, as an early indicator or avoidable premature obsolescence.

The *full* system of building regulation and governance needs to be mapped and considered to accelerate the scale and speed of building retrofits for both residential and non-residential buildings. Interventions need to be viewed and designed so that they mutually reinforce each other. Both positive and negative, coercive and persuasive interventions by local and national governments need to be considered simultaneously and alongside the roles that NGOs and market-based incentives contribute in making effective impacts upon retrofitting existing buildings. Governments can take up the role of the 'orchestrator' of these interventions and make sure that a broad variety of interventions are attractive to as large a group of property developers, owners, and tenants as possible, from the early movers to the laggards late to adopting retrofitting to meet climate emergency and make a positive contribution to our overall well-being.

Policy safeguards need to be built into any future disaster recovery plans to avoid public money being invested into socially or environmentally harmful activities that could worsen current and future crises, negatively impact human health and social equity, and are not at least neutral in addressing the climate emergencies. This involves aligning any public financing of retrofitting activity with climate pathways and taking a holistic and comprehensive approach to retrofitting buildings rather than single-element measures to optimise reduction in energy consumption and human health.

Development of policy instruments to capture and publish whole lifecycle data useful for future retrofit decisions for residential and non-residential real estate assets at the point of construction, sale, or advertising for lease is needed. It is also necessary to design and effectively communicate transparent policy instruments that support the financing of retrofitting of residential buildings so that access to these schemes can be navigated easily by residential owners and landlords, and have straightforward processes for the applicants to access the finance. Equity of access to retrofit finance is also an important factor to build into policy instruments.

10.5.2 A manifesto of recommendations for educationalists

Education is vital. When students are introduced to best practice concepts in sustainability, resilience, circular economy, and retrofitting early in their schooling from primary to tertiary education, they become embedded and familiar. Learning behaviours early tend to stick throughout a lifetime. The shock of the new is a phenomenon, whereby people find it hard to accept new ways of thinking and to change their actions and behaviours to accommodate new thinking.

Education departments have a crucial role to play in school design and maintenance. They can show children renewable energy technologies, reuse and recycling best practices, composting and urban food production.

Teachers, likewise, through the curriculum, can introduce concepts of reducing waste, lowering energy and water consumption, and using natural and biomaterials wherever possible from the start. These ideas can be taught in a diverse range of subjects such as science, biology, geography, technology, politics, English, and applied maths. Exercises that enable students to apply concepts in practical projects also build understanding and capacity.

The tertiary sector should continue to develop courses and assignments that expose students to the latest thinking, technologies and applications of sustainability and resilience across the built environment, from planning, construction, property, law, and engineering. Increased options of trans-disciplinary subjects with science and engineering faculties should be encouraged to further develop knowledge and

Table 10.1 The resilient building retrofit manifesto

Policymakers
1. Access to fine-grain data to understand vacancy, as an early indicator of premature obsolescence and identify the need to adapt existing buildings.
2. Oversight and deep understanding of the full system of building regulation and governance, so that policy interventions mutually reinforce each other to appeal to the broadest range of property owners and developers from early adopters to the laggards.
3. Policy safeguards need to be built into any future disaster recovery plans to avoid public money being invested into socially or environmentally harmful activities.
4. Development of transparent policy instruments to increase retrofit uptake, coupled with effective communication strategies and a comprehensive approach to retrofitting buildings rather than single element measures.

Educationalists
1. Education is vital in framing the urgency and communicating best practice concepts in sustainability, resilience, circular economy and retrofitting.
2. Education departments have a crucial role to play – we must change thinking: mainstream retrofitting as the primary go-to strategy for development.
3. Teachers through the curriculum play a crucial role.
4. The tertiary sector must be the seat of innovation and teach people how to apply best practices.

Professional bodies
1. Must build capacity in knowledge and skills for new and existing members through CPD and course accreditations.
2. Must ensure resilience and sustainability principles are a core characteristic of all members.
3. Must fund and support innovation and development of new best practices.
4. Must use their influence to advocate wherever possible for greater standards in respect of resilient building retrofit.

Practitioners
1. Should find common ground and come together across disciplines to share knowledge and advance retrofitting practice.
2. Need to adopt a retrofit-first approach to development.
3. Need to upskill to take a holistic lifecycle analysis approach to minimise the environmental impact of development decisions.
4. Must adhere to testing any newly installed building services and designs to ensure that they are functioning correctly and operate as intended.

understanding in bio technologies and hi-tech approaches to optimise building performance. The educationalists contributions are summarised in Table 10.1.

10.5.3 A manifesto of recommendations for professional bodies

Professional bodies exist in all countries to provide best practices and standards in their members. Some professional bodies such as the Royal Institution of Chartered Surveyors (RICS) are global. Professional bodies in the built environment include the following disciplines: property, valuation, facility management, construction or building, quantity surveyors, urban and town planning, engineering (mechanical, civil, and structural).

The professional bodies provide minimum standards for entry and options to attain a more senior level of membership to reflect a member's experience and expertise. Typically, new members can seek to join a body having completed an accredited degree and a minimum of two years structured work experience. After this period, documents outlining their experience are submitted and an assessment of professional competence by a panel of peers is undertaken. Once membership is attained, there is a commitment to ongoing professional development and education through a minimum number of hours each year. This is known as continuing professional development or CPD and takes the form of reading and attending CPD events hosted by the professional body. Professional bodies have the capacity to ensure resilience and sustainability principles are a core characteristic of all members.

The professional bodies fund and support innovation and development of new ideas and practices for their members. They develop technical guides and materials to support members learning and ability to apply new knowledge in their professional practices. The professional bodies have a moral and ethical duty to ensure that members fully understand all the relevant issues relating to resilient building retrofit, through the provision of materials and best practice guides. They have a duty to ensure that the professional development courses they accredit also contain current materials and apply contemporary best practices and thinking towards resilient building retrofits.

Professional bodies can also influence policymakers and government, and importantly their clients through the prestige and status of their membership across the areas of practice. They must exercise this power to encourage policymakers, owners, and clients to adopt best practices to reduce and minimise environmental impact, to adopt best practices in sustainability and resilience in building retrofits. The professional body contributions are summarised in Table 10.1.

10.5.4 A manifesto of recommendations for practitioners

Sectoral fragmentation between practitioners needs to be addressed by practitioners as fragmentation results in a multitude of disconnections and reduces the quality of information available to make retrofitting decisions.

In practice, there should be a commitment by all professionals for a retrofit-first approach to managing the built environment. This includes all building elements, taking a holistic lifecycle analysis approach to minimise the environmental impact and ensure human health is not impacted. Measures to prioritise are measures to achieve higher levels of energy efficiency and reduce risks such as seismic strengthening, ensuring weather tightening, and providing social equity through disability access, and amenities to support active transport modes.

To develop a whole lifecycle approach, coordination of information by practitioners, its storage and its handover to owners are crucial, as well as a commitment by practitioners to address gaps in information for future retrofitting decisions. There is an urgency for practitioners to develop and adopt a universal, standardised system or protocol to store, access, update and transfer building-related data and information to assist retrofitting decisions. On top of this, critical thinking and research literacy skills are crucial for practitioners to develop so that robust data can be identified during the design stages, particularly when evaluating new or innovative applications of technology and materials in retrofit designs. Persuasion may be needed to adopt a retrofit-first approach through evidencing the broad range of benefits to nudge client behaviours when dealing with laggard owners and owner groups.

Practitioners adhere to testing and adjusting any newly installed building services to ensure they are functioning correctly prior to formal handover from the project team. This is vital given that many new buildings, and retrofitted buildings, in practice, fail to operate as designed. Practitioners need to be mindful of the different kinds of ownership groups of existing buildings who are potential clients making decisions to retrofit residential buildings, such as owner-occupiers, landlords and buy-to-let sole investors and investor groups. For non-residential buildings, there are owner-occupiers, mum and dad investors, and small to large property investment firms.

10.6 Conclusions

Several themes cut across the chapters in this book. First is the need for action and the kinds of actions that could radically improve the take-up of retrofitting and adaptive reuse across the different sectors which are discussed: from residential and non-residential buildings such as office, retail and industrial buildings; in heritage-listed buildings and buildings whose heritage value is not yet widely recognised; and for understanding buildings stocks and for proposing effective action.

A second theme is a need for the collection of data and its coordination, storage and accessibility to inform approaches for sustainably managing built environment assets and for developing policy mechanisms to aid retrofitting and adaptive reuse. Despite living in the 'information age', where the term 'smart' city or building is familiar to many, access to data is relatively poor, particularly data to aid policy and practice in sustainable retrofitting and adaptive reuse of buildings at risk of premature obsolescence or buildings which do not have adequate performance for the occupants.

Thirdly, the need for holistic approaches cuts across many of the chapters, coupled with the need to exchange knowledge between end-users and professionals, to achieve best practice, extending and minimising environmental impact at all different stages of a building's life cycle. The need for holistic thinking also cuts across financial incentives and public expenditure, which seeks to mitigate the acute stresses cities face now and in the future. Holistic thinking around financial incentives must ensure public spending is not diverted into built environment development, which locks in highly polluting infrastructure, potentially worsening the current but also any future crises.

The final cross-cutting theme of this book is the need to understand building regulation and governance as a complex system and to critically reflect when building regulation and policy instruments are routinely cited as a barrier to retrofitting and adaptive reuse. What evidence is presented and which specific aspects of policy, regulation and governance are problematic and why? That is not to say there are no problems to overcome, far from it. However, it is easy to use simplistic, broad-brush claims about 'red-tape' to avoid adopting sustainable, innovative or new practices when systems are complex. The challenges need unpacking further by all stakeholders to the continued advancement of retrofitting knowledge continues. Effective monitoring of policy measures brought in to enable retrofitting and adaption interventions ensure unintended consequences of policy measures are mitigated swiftly where consequences have environmental and social impacts. The research on which each chapter draws highlights the urgency for policymakers, educationalists, professional bodies, and practitioners alike to mainstream a retrofit and adaptive reuse-first approach to built environment development. Practical frameworks, both newly presented and international examples, are offered to act as a gateway to change. Building and technology case studies contained in this text signpost the reader to completed successful examples of adaptations and applications of technologies, highlighting a wide range of realisable possibilities. Our manifesto draws upon these and encourages the reader to read further using these case studies as a starting point.

References

Buitelaar, E., Moroni, S. and De Franco, A., 2021. Building obsolescence in the evolving city. Reframing property vacancy and abandonment in the light of urban dynamics and complexity. *Cities*, 108(2021), p. 102964. https://doi.org/10.1016/j.cities.2020.102964.

Clayton, J., Devaney, S., Sayce, S. and van de Wetering, J., 2021. *Climate Risk and Commercial Property Values: A Review and Analysis of the Literature*. UNEP FI. Available from www.unepfi.org/wordpress/wp-content/uploads/2021/08/Climate-risk-and-real-estate-value_Aug2021.pdf. Accessed 9 February 2022.

Index

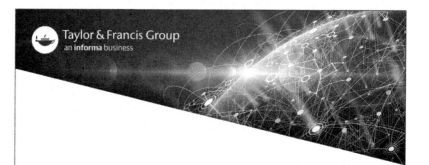

Printed in the United States
by Baker & Taylor Publisher Services